Nonlinear Optimization with Financial Applications

Nonlinear Optimization with Financial Applications

Michael Bartholomew-Biggs
University of Hertfordshire
Hatfield, England

KLUWER ACADEMIC PUBLISHERS
Boston / Dordrecht / London

Library of Congress Cataloging-in-Publication Data

A C.I.P. Catalogue record for this book is available
from the Library of Congress.

ISBN 978-1-4899-8119-6 ISBN 978-0-306-48670-8 (eBook)

Softcover re-print of the Hardcover 1st edition 2007

9 8 7 6 5 4 3 2 1 SPIN 11161363

springeronline.com

Contents

List of Figures

List of Figures

List of Tables

Preface

This book has grown out of undergraduate and postgraduate lecture courses given at the University of Hertfordshire and the University of Bergamo. Its primary focus is on numerical methods for nonlinear optimization. Such methods can be applied to many practical problems in science, engineering and management: but, to provide a coherent secondary theme, the applications considered here are all drawn from financial mathematics. (This puts the book in good company since many classical texts in mathematics also dealt with commercial arithmetic.) In particular, the examples and case studies are concerned with portfolio selection and with time-series problems such as fitting trend-lines and trend-channels to market data.

The content is intended to be suitable for final-year undergraduate students in mathematics (or other subjects with a high mathematical or computational content) and exercises are provided at the end of most sections. However the book should also be useful for postgraduate students and for other researchers and practitioners who need a foundation for work involving development or application of optimization algorithms. It is assumed that readers have an understanding of the algebra of matrices and vectors and of the Taylor and Mean Value Theorems in several variables. Prior experience of using computational methods for problems such as solving systems of linear equations is also desirable, as is familiarity with iterative algorithms (e.g., Newton's method for nonlinear equations in one variable).

The approach adopted in this book is a blend of the practical and theoretical. A description and derivation is given for most of the currently popular methods for continuous nonlinear optimization. For each method, important convergence results are outlined (and we provide proofs when it seems instructive to do so). This theoretical material is complemented by numerical illustrations which give a flavour of how the methods perform in practice.

It is not always obvious where to draw the line between general descriptions of algorithms and the more subtle and detailed considerations relating to research issues. The particular themes and emphases in this book have grown out of the author's experience at the Numerical Optimization Centre (NOC). This was established in 1968 at the Hatfield College of Technology (predecessor of the University of Hertfordshire) as a centre for research in optimization techniques. Since its foundation, the NOC has been engaged in algorithm development and consultancy work (mostly in the aerospace industry). The NOC staff has included, at various times, Laurence Dixon, Ed Hersom, Joanna Gomulka, Sean McKeown and Zohair Maany who have all made contributions to the state-of-the-art in fields as diverse as quasi-Newton methods, sequential quadratic programming, nonlinear least-squares, global optimization, optimal control and automatic differentiation.

The computational results quoted in this book have been obtained using a Fortran90 module called `SAMPO`. This is based on the NOC's `OPTIMA` library – a suite of subroutines for different types of minimization problem. The name `SAMPO` serves as an acronym for Software And Methods for Portfolio Optimization. (However, it is also worth mentioning that *The Sampo* appears in Finnish mythology as a magical machine which grinds out endless supplies of corn, salt and gold. Its relevance to a book about financial mathematics needs no further comment.) The `SAMPO` software is not described in detail in this book (which does not deal with any programming issues). Interested readers can obtain it from an `ftp` site as specified in the appendix. Some of the student exercises can be attempted using `SAMPO` but most of them can also be tackled in other ways. For instance, the `SOLVER` tool in Microsoft Excel can handle both constrained and unconstrained optimization problems. Alternatively, users of MATLAB, can access a comprehensive toolbox of optimization procedures. The NAG libraries in C and Fortran include a wide range of minimization routines and the NEOS facility at the Argonne National Laboratories allows users to submit optimization problems via email.

I am indebted to a number of people for help in the writing of this book. Besides the NOC colleagues already mentioned, I would like to thank Alan Davies and all the staff in the Mathematics Department at the University of Hertfordshire for their support. I am particularly grateful to Don Busolini and Steve Kane, for introducing me to the financial applications in this book, and to Steve Parkhurst for sharing the lecture courses which underpin it. I have received encouragement and advice from Marida Bertocchi of the Universty of Bergamo, Alistair Forbes of the National Physical Laboratory, Berc Rustem of

Imperial College and Ming Zuo of the University of Alberta. Useful comments and criticisms have also been made by students who were the guinea-pigs for early versions of the text. Surrounded by such clouds of witnesses, any mistakes or omissions that remain in the text are entirely my responsibility.

My thanks are also due to John Martindale and Angela Quilici-Burke at Kluwer Academic Publishing for their encouragement and help with the preparation of the final version of the book.

Most of all, my deepest thanks go to my wife Nancy Mattson who put up with the hours I spent *incommunicado* and crouched over my laptop! Nancy does not share my attachment to computational mathematics: but she and I do have a common commitment to poetry. Therefore, scattered through the book, the reader will find a number of short mathematically-flavoured poems which occupy pages that would otherwise have been mere white space. To those who have asked how mathematical aptitude interacts with poetic composition a reply could be

Two cultures

Poets show, don't tell:
build metaphors from concrete
and specific bricks.

In mathematics
the abstract and general's
our bread and butter.

For more on this subject, the reader may wish to consult [66, 67, 68].

Chapter 1

PORTFOLIO OPTIMIZATION

1. Nonlinear optimization

Optimization involves finding minimum (or maximum) values of functions. Practical applications include calculating a spacecraft launch trajectory to maximize payload in orbit or planning distribution schedules to minimize transportation costs or delivery times. In such cases we seek values for certain *optimization variables* to obtain the *optimum* value for an *objective function*

We now give a simple example of the formulation of an optimization problem. Suppose we have a sum of money M to split between three managed investment funds which claim to offer percentage rates of return r_1, r_2 and r_3. If we invest amounts y_1, y_2 and y_3 we can expect our overall return to be

$$R = \frac{r_1 y_1 + r_2 y_2 + r_3 y_3}{M} \%.$$

If the management charge associated with the i-th fund is calculated as $c_i y_i$, then the total cost of making the investment is

$$C = c_1 y_1 + c_2 y_2 + c_3 y_3.$$

Suppose we are aiming for a return $R_p\%$ and that we want to pay the least charges to achieve this. Then we need to find y_1, y_2 and y_3 to solve the problem

$$\text{Minimize} \quad c_1 y_1 + c_2 y_2 + c_3 y_3 \tag{1.1.1}$$

$$\text{subject to} \quad r_1 y_1 + r_2 y_2 + r_3 y_3 = MR_p, \quad y_1 + y_2 + y_3 = M \tag{1.1.2}$$

$$\text{and} \quad y_1 \geq 0,\ y_2 \geq 0,\ y_3 \geq 0. \tag{1.1.3}$$

This is an optimization problem involving both equality and inequality *constraints* to restrict the values of the variables. The inequalities are included because investments must obviously be positive. If we tried to solve the problem without these restrictions then an optimization algorithm would attempt to reduce costs by making one or more of the y_i large and negative.

In fact, since they only involve linear expressions, (1.1.1) – (1.1.3) represents a *Linear Programming* problem. However, we do not wish to limit ourselves to linear programming (which is a substantial body of study in its own right – see, for instance, [1] – and involves a number of specialised methods not covered in this book). Instead, we shall be concerned with the more general problem in which the function and/or the constraints are nonlinear.

The simple investment decision considered above could be represented as a *nonlinear* programming problem if we imagined that an attempt to make a negative investment would be penalised by a very high management charge! Thus we could consider the problem

$$\text{Minimize} \quad c_1y_1 + c_2y_2 + c_3y_3 + K\sum_{i=1}^{3}\psi(y_i) \tag{1.1.4}$$

$$\text{subject to} \quad r_1y_1 + r_2y_2 + r_3y_3 = MR_p \quad \text{and} \quad y_1 + y_2 + y_3 = M. \tag{1.1.5}$$

where K is a large positive constant and the function ψ is defined by

$$\psi(x) = \begin{cases} 0 & \text{if } x \geq 0 \\ x^2 & \text{if } x < 0 \end{cases}$$

The objective function (1.1.4) is now nonlinear (being linear for some values of the y_i and quadratic for others) and it features two linear equality constraints.

Equality constraints can sometimes be used to eliminate variables from an optimization problem. Thus, since $y_3 = M - y_1 - y_2$, we can transform (1.1.4), (1.1.5) into the problem of minimizing the two-variable function

$$c_1y_1 + c_2y_2 + c_3(M - y_1 - y_2) + K\sum_{i=1}^{2}\psi(y_i) + K\psi(M - y_1 - y_2) \tag{1.1.6}$$

subject to the single constraint

$$r_1y_1 + r_2y_2 + r_3(M - y_1 - y_2) = MR_p. \tag{1.1.7}$$

Obviously (see Exercise 1 below) we could go further and use the equality constraint (1.1.7) to express y_1 in terms of y_2. Our problem could then be expressed as an *unconstrained* minimization of a function of a single variable.

In the chapters which follow we shall describe several methods for solving nonlinear optimzation problems, with and without constraints. Applications of these methods will be taken from the field of portfolio optimization, using ideas due to Markowitz [2, 3] which are introduced in the next section.

Exercises

1. Reformulate (1.1.6), (1.1.7) to give an unconstrained minimization problem involving y_1 only.

2. Using the values $M = 1000, R_p = 1.25, K = 10^{-3}$ together with

$$c_1 = 0.01, \quad c_2 = 0.009, \quad c_3 = 0.014 \quad \text{and} \quad r_1 = 1.2, \quad r_2 = 1.1, \quad r_3 = 1.4$$

plot a graph of the function obtained in question 1 in the range $0 \leq y_i \leq M$. Hence deduce the minimum-cost that produces a return of 1.25%.
How does the solution change in the cases where $c_3 = 0.012$ and $c_3 = 0.011$?

3. Formulate the problem of finding the maximum return that can be obtained for a fixed management cost C_f. (Note that the problem of maximizing a function $F(x)$ is equivalent to the problem of minimizing $-F(x)$.)

2. Portfolio return and risk

Suppose we have a history of percentage returns, over m time periods, for each of a group of n assets (such as shares, bonds etc.). We can use this information as a guide to future investments. As an example, consider the following data for three assets over six months.

	Return % for					
	January	February	March	April	May	June
Asset 1	1.2	1.3	1.4	1.5	1.1	1.2
Asset 2	1.3	1.0	0.8	0.9	1.4	1.3
Asset 3	0.9	1.1	1.0	1.1	1.1	1.3

Table 1.1. Monthly rates of return on three assets

In general, we can calculate the mean return $\bar{r}_i$ for each asset as

$$\bar{r}_i = \frac{\sum_{j=1}^{m} r_{ij}}{m}.$$

where r_{ij} $(i = 1,..,n, \ j = 1,..,m)$ denotes the return on asset i in period j. Hence, for the data in Table 1.1 we get

$$\bar{r}_1 \approx 1.2833; \quad \bar{r}_2 \approx 1.1167; \quad \bar{r}_3 \approx 1.0833. \qquad (1.2.1)$$

If we spread an investment across the n assets and if y_i denotes the fraction invested in asset i then the values of the y_i define a *portfolio*. Since *all* investment must be split between the n assets, the invested fractions must satisfy

$$S = \sum_{i=1}^{n} y_i = 1. \tag{1.2.2}$$

The *expected portfolio return* is given by

$$R = \sum_{i=1}^{n} \bar{r}_i y_i. \tag{1.2.3}$$

Thus, for the data in Table 1.1, we might choose to put half our investment in asset 1, one-third in asset 2 and one-sixth in asset 3. This would give an expected return

$$R = 1.2833 \times \frac{1}{2} + 1.1167 \times \frac{1}{3} + 1.0833 \times \frac{1}{6} \approx 1.194\%.$$

The *risk* associated with a particular portfolio is determined from *variances and covariances* that can be calculated from the history of returns r_{ij}. The variance of asset i is

$$\sigma_i^2 = \frac{\sum_{j=1}^{m} (r_{ij} - \bar{r}_i)^2}{m} \tag{1.2.4}$$

while the covariance of assets i and k is

$$\sigma_{ik} = \frac{\sum_{j=1}^{m} (r_{ij} - \bar{r}_i)(r_{kj} - \bar{r}_k)}{m}. \tag{1.2.5}$$

Evaluating (1.2.4), (1.2.5) for the returns in Table 1.1 we get

$$\sigma_1^2 = 0.0181; \ \ \sigma_2^2 = 0.0514; \ \ \sigma_3^2 = 0.0147; \tag{1.2.6}$$

$$\sigma_{12} = -0.0281; \ \ \sigma_{13} = -0.00194; \ \ \sigma_{23} = 0.00528. \tag{1.2.7}$$

The *variance* of the portfolio defined by the investment fractions $y_1, ..., y_n$ is

$$V = \sum_{i=1}^{n} \sigma_i^2 y_i^2 + 2 \sum_{i=1}^{n-1} \sum_{j=i+1}^{n} \sigma_{ij} y_i y_j \tag{1.2.8}$$

which can be used as a measure of portfolio risk. Thus for a three asset problem

$$V = \sigma_1^2 y_1^2 + \sigma_2^2 y_2^2 + \sigma_3^2 y_3^2 + 2\sigma_{12} y_1 y_2 + 2\sigma_{13} y_1 y_3 + 2\sigma_{23} y_2 y_3. \tag{1.2.9}$$

Using (1.2.6) and (1.2.7), the risk function V for the data in Table 1.1 is

$$0.0181y_1^2 + 0.0514y_2^2 + 0.0147y_3^2 - 0.0562y_1y_2 - 0.00388y_1y_3 + 0.01056y_2y_3.$$

Thus, for a portfolio defined by $y_1 = \frac{1}{2}$, $y_2 = \frac{1}{3}$, $y_3 = \frac{1}{6}$

$$V = 0.0181 \times \frac{1}{4} + 0.0514 \times \frac{1}{9} + 0.0147 \times \frac{1}{36} - 0.0563 \times \frac{1}{6}$$

$$-0.00388 \times \frac{1}{12} + 0.01056 \times \frac{1}{18}$$

which gives a risk $V \approx 0.00153$.

The return and risk functions (1.2.3) and (1.2.8) can be written more conveniently using matrix-vector notation. Expression (1.2.3) is equivalent to

$$R = \bar{r}^T y \qquad (1.2.10)$$

where $\bar{r}$ denotes the column vector $(\bar{r}_1, \bar{r}_2, ..., \bar{r}_n)^T$ and y is the column vector of invested fractions $(y_1, y_2, ..., y_n)^T$. Moreover, we can express (1.2.8) as

$$V = y^T Q y \qquad (1.2.11)$$

where the *variance-covariance matrix* Q is defined by

$$Q = \begin{pmatrix} \sigma_{11} & \sigma_{12} & \cdots & \sigma_{in} \\ \sigma_{12} & \sigma_{22} & \cdots & \sigma_{2n} \\ \cdots & \cdots & \cdots & \cdots \\ \cdots & \sigma_{ij} & \cdots & \cdots \\ \cdots & \cdots & \cdots & \cdots \\ \sigma_{1n} & \sigma_{2n} & \cdots & \sigma_{nn} \end{pmatrix}. \qquad (1.2.12)$$

In (1.2.12) we have used the equivalent notation

$$\sigma_{ii} = \sigma_i^2. \qquad (1.2.13)$$

Thus, for data in Table 1.1,

$$V = \begin{pmatrix} y_1 & y_2 & y_3 \end{pmatrix} \begin{pmatrix} 0.0181 & -0.0281 & -0.00194 \\ -0.0281 & 0.0514 & 0.00528 \\ -0.00194 & 0.00528 & 0.0147 \end{pmatrix} \begin{pmatrix} y_1 \\ y_2 \\ y_3 \end{pmatrix}.$$

We can also write (1.2.2) in vector form as

$$S = e^T y = 1 \qquad (1.2.14)$$

where e denotes the n-vector $(1, 1, .., 1)^T$.

Exercises

1. Returns on assets such as shares can be obtained from day-to-day stock market prices. If the closing prices of a share over five days trading are

$$P_1 = 4.62,\ P_2 = 4.39,\ P_3 = 3.64,\ P_4 = 3.89,\ P_5 = 4.23$$

calculate the corresponding returns using the formula

$$r_i = 100\frac{P_i - P_{i-1}}{P_{i-1}}.$$

What returns are given by the alternative formula

$$r_i = 100\log_e(\frac{P_i}{P_{i-1}})?$$

Show that the two formulae for r_i give similar results when $|P_i - P_{i-1}|$ is small.

2. Calculate the mean returns and the variance-covariance matrix for the asset data in Table 1.2.

	Return % for				
	Day 1	Day 2	Day 3	Day 4	Day 5
Asset 1	0.15	0.26	0.18	0.04	0.06
Asset 2	0.04	-0.07	-0.05	0.07	0.03
Asset 3	0.11	0.21	0.06	-0.06	0.12

Table 1.2. Daily rates of return on three assets

3. Optimizing two-asset portfolios

We begin by considering simple portfolios involving only two assets. There are several ways to define an optimal choice for the invested fractions y_1 and y_2 and each of them leads to a one-variable minimization problem. The ideas introduced in this section can be extended to the more realistic case when there are n (> 2) assets.

The basic minimum risk problem

A major concern in portfolio selection is the minimization of risk. In its simplest form, this means finding invested fractions $y_1, ..., y_n$ to solve the problem

Minrisk0

$$\text{Minimize } V = y^T Q y \quad \text{subject to } \sum_{i=1}^{n} y_i = 1. \tag{1.3.1}$$

Like the earlier example (1.1.4), (1.1.5), this is an equality constrained minimization problem for which a number of solution methods will be given in later chapters. Our first approach involves using the constraint to eliminate y_n and yield an unconstrained minimization problem involving only $y_1, \ldots, y_{n-1}$.

In the particular case of a two-asset problem, we wish to minimize $V = y^T Qy$ with the constraint $y_1 + y_2 = 1$. If we write x in place of the unknown fraction y_1 then $y_2 = 1 - y_1 = 1 - x$. We can now define

$$\alpha = \begin{pmatrix} 0 \\ 1 \end{pmatrix} \quad \beta = \begin{pmatrix} 1 \\ -1 \end{pmatrix} \tag{1.3.2}$$

and then it follows that $y = \alpha + \beta x$ so that problem **Minrisk0** becomes

$$\text{Minimize} \quad V = (\alpha + \beta x)^T Q(\alpha + \beta x). \tag{1.3.3}$$

Expanding the matrix-vector product we get

$$V = \alpha^T Q\alpha + (2\beta^T Q\alpha)x + (\beta^T Q\beta)x^2.$$

At a minimum, the first derivative of V is zero. Now

$$\frac{dV}{dx} = 2\beta^T Q\alpha + 2\beta^T Q\beta x \tag{1.3.4}$$

and so there is a stationary point at

$$x = -\frac{\beta^T Q\alpha}{\beta^T Q\beta}. \tag{1.3.5}$$

This stationary point will be a minimum (see chapter 2) if the second derivative of V is positive. In fact

$$\frac{d^2V}{dx^2} = 2\beta^T Q\beta$$

and we shall be able to show later on that this cannot be negative.

We now consider a numerical example based on the following Table 1.3.

	Return % for					
	January	February	March	April	May	June
Asset 1	1.2	1.3	1.4	1.5	1.1	1.2
Asset 2	1.3	1.0	0.8	0.9	1.4	1.3

Table 1.3. Monthly rates of return on two assets

These are, in fact, simply the returns for the first two assets in Table 1.1 and so $\bar{r}$ is given by (1.2.1) and the elements of Q come from (1.2.6) and (1.2.7). Specifically we have

$$\bar{r} = \begin{pmatrix} 1.2833 \\ 1.1167 \end{pmatrix} \quad \text{and} \quad Q = \begin{pmatrix} 0.0181 & -0.0281 \\ -0.0281 & 0.0514 \end{pmatrix}. \tag{1.3.6}$$

If we use Q from (1.3.6) then (1.3.3) becomes

$$V = 0.0514 - 0.1590x + 0.1256x^2.$$

Hence, by (1.3.4) and (1.3.5),

$$\frac{dV}{dx} = -0.159 + 0.2512x \quad \text{and so} \quad x = \frac{0.159}{0.2512} \approx 0.633.$$

Since V has a positive second derivative,

$$\frac{d^2V}{dx^2} = 0.2512,$$

we know that a minimum of V has been found. Hence the minimum-risk portfolio for the assets in Table 1.3 has $y_1 = x \approx 0.633$. Obviously the invested fraction $y_2 = 1 - y_1 \approx 0.367$ and so the "least risky" strategy is to invest about 63% of the capital in asset 1 and 37% in asset 2. The portfolio risk is then $V \approx 0.00112$ and, using the $\bar{r}$ values in (1.3.6), the expected portfolio return is

$$R \approx 0.633 \times \bar{r}_1 + 0.367 \times \bar{r}_2 \approx 1.22\%.$$

Exercise
Find the minimum-risk portfolio involving the first two assets in Table 1.2.

Optimizing return and risk

The solution to problem **Minrisk0** can sometimes be useful; but in practice we will normally be interested in both risk and return rather than risk on its own.

In a rather general way we can say that an optimal portfolio is one which gives "biggest return at lowest risk". One way of trying to determine such a portfolio is to consider a composite function such as

$$F = -R + \rho V = -\bar{r}^T y + \rho y^T Q y. \tag{1.3.7}$$

The first term is the negative of the expected return and the second term is a multiple of the risk. If we choose invested fractions y_i to minimize F then we can expect to obtain a *large* value for return coupled with a *small* value for risk. The positive constant ρ controls the balance between return and risk.

Based on the preceding discussion, we now consider the problem

Risk-Ret1

$$\text{Minimize} \quad F = -\bar{r}^T y + \rho y^T Q y \quad \text{subject to} \quad \sum_{i=1}^{n} y_i = 1. \tag{1.3.8}$$

As in the previous section, we can use the equality constraint to eliminate y_n and then find the unconstrained minimum of F, considered as a function of $y_1, ..., y_{n-1}$. In particular, for the two-asset case, we can we write x in place of the unknown y_1 and define α and β by (1.3.2) so that $y = \alpha + \beta x$. Problem **Risk-Ret1** then becomes

$$\text{Minimize } F = -\bar{r}^T \alpha - (\bar{r}^T \beta)x + \rho[\alpha^T Q \alpha + 2(\beta^T Q \alpha)x + (\beta^T Q \beta)x^2]. \tag{1.3.9}$$

Differentiating, we get

$$\frac{dF}{dx} = -\bar{r}^T \beta + 2\rho\beta^T Q\alpha + 2\rho(\beta^T Q\beta)x$$

and so a stationary point of F occurs at

$$x = \frac{\bar{r}^T \beta - 2\rho\beta^T Q\alpha}{2\rho(\beta^T Q\beta)}. \tag{1.3.10}$$

We now consider the data in Table 1.3. As before, values for $\bar{r}$ and Q are given by (1.3.6) and so F becomes

$$F = 0.1256\rho x^2 - (0.1667 + 0.1589\rho)x - 1.1167 + 0.05139\rho.$$

In order to minimize F we solve

$$\frac{dF}{dx} = 0.2512\rho x - (0.1667 + 0.1589\rho) = 0.$$

This gives

$$x = \frac{0.6636}{\rho} + 0.6326 \tag{1.3.11}$$

and this stationary point is a minimum because the second derivative of F is 0.2512ρ which is positive whenever $\rho > 0$.

For $\rho = 5$, (1.3.11) gives $y_1 = x \approx 0.765$. This implies $y_2 \approx 0.235$ and the portfolio expected return is about 1.244% with risk $V \approx 0.00333$. If we choose $\rho = 10$ (thus placing more emphasis on reducing risk) the optimal portfolio is

$$y_1 \approx 0.7, \; y_2 \approx 0.3 \quad \text{giving} \quad R \approx 1.233\% \; \text{ and } \; V \approx 0.00167.$$

Exercises

1. Solve (1.3.9) using data for assets 2 and 3 in Table 1.2 and taking $\rho = 100$.

2. Based on the data in Table 1.3, determine the range of values of ρ for which the solution of (1.3.9) gives x lying between 0 and 1. What, *in general*, is the range for ρ which ensures that (1.3.10) gives $0 \le x \le 1$?

Minimum risk for specified return

Problem **Risk-Ret1** allows us to balance risk and return according to the choice of the parameter ρ. Another approach could be to fix a target value for return, say R_p%, and to consider the problem

Minrisk1

$$\text{Minimize} \quad V = y^T Q y \tag{1.3.12}$$

$$\text{subject to} \quad \sum_{i=1}^{n} \bar{r}_i y_i = R_p \quad \text{and} \quad \sum_{i=1}^{n} y_i = 1. \tag{1.3.13}$$

One way to tackle **Minrisk1** is to consider the *modified* problem

Minrisk1m

$$\text{Minimize} \quad F = y^T Q y + \frac{\rho}{R_p^2}\left(\sum_{i=1}^{n} \bar{r}_i y_i - R_p\right)^2 \quad \text{subject to} \quad \sum_{i=1}^{n} y_i = 1. \tag{1.3.14}$$

At a minimum of F we can expect the risk $y^T Q y$ to be small and also that the return $\bar{r}^T y$ will be close to the target figure R_p. As in **Risk-Ret1**, the value of the parameter ρ will determine the balance between return and risk.

For the two-asset case we can solve **Minrisk1m** by eliminating y_2 and using the transformation $y = \alpha + \beta x$ where $x = y_1$. We then get the problem

$$\text{Minimize} \quad F = (\alpha + \beta x)^T Q (\alpha + \beta x) + \frac{\rho}{R_p^2}(\bar{r}^T \alpha - R_p + \bar{r}^T \beta x)^2. \tag{1.3.15}$$

After some simplification, and writing $\bar{\rho}$ in place of ρ / R_p^2, F becomes

$$\alpha^T Q \alpha + \bar{\rho}(\bar{r}^T \alpha - R_p)^2 + 2(\beta^T Q \alpha + \bar{\rho}(\bar{r}^T \alpha - R_p)\bar{r}^T \beta)x + (\beta^T Q \beta + \bar{\rho}(\bar{r}^T \beta)^2)x^2.$$

A minimum of F occurs when its first derivative is zero, i.e. at

$$x = -\frac{\beta^T Q \alpha + \bar{\rho}(\bar{r}^T \alpha - R_p)\bar{r}^T \beta}{\beta^T Q \beta + \bar{\rho}(\bar{r}^T \beta)^2} \tag{1.3.16}$$

For the assets in Table 1.3, values of $\bar{r}$ and Q are given in (1.3.6) and so

$$\bar{r}^T \alpha \approx 1.1167; \quad \bar{r}^T \beta \approx 0.1667; \quad \beta^T Q \alpha \approx -0.0794; \quad \beta^T Q \beta \approx 0.1256.$$

Hence (1.3.16) gives

$$x \approx -\frac{-0.0794 + 0.1667\bar{\rho}(1.1167 - R_p)}{0.1256 + 0.1667^2\bar{\rho}}.$$

If the target return R_p is 1.25% then (1.3.16) becomes

$$x \approx -\frac{-0.0794 - 0.0222\bar{\rho}}{0.1256 + 0.0278\bar{\rho}} \approx -\frac{-0.0794 - 0.0142\rho}{0.1256 + 0.0178\rho}.$$

Setting $\rho = 10$ gives $x \approx 0.2214/0.3036 \approx 0.73$. Hence the corresponding invested fractions are

$$y_1 \approx 0.73, \quad y_2 \approx 0.27$$

giving portfolio risk and return $V \approx 0.00233$ and $R \approx 1.24\%$.

Solutions with $\bar{r}^T y$ closer to its target value can be obtained by increasing the parameter ρ. Thus, when $\rho = 100$, we get the solution

$$y_1 \approx 0.79, \quad y_2 \approx 0.21 \quad \text{with} \quad y^T Q y \approx 0.0042 \quad \text{and} \quad \bar{r}^T y \approx 1.25\%.$$

Exercises

1. For the data in Table 1.3, estimate the minimum-risk portfolio for a target return of 1.2% using values of $\rho = 10, 100, 1000$ in the function (1.3.15).

2. If, for a two-asset problem, R_p is taken as $0.5(\bar{r}_1 + \bar{r}_2)$, is it necessarily true that the solution to **Minrisk1m** has y_1 and y_2 both approximately equal to 0.5? What can you say about the possible ranges for y_1 and y_2 when ρ is large?

4. Minimimum risk for three-asset portfolios

In the case of a three-asset portfolio we can reduce problem **Minrisk1** to a one-variable minimization by using *both* constraints to eliminate the invested fractions y_2 and y_3. Equations (1.3.13) imply

$$\bar{r}_2 y_2 + \bar{r}_3 y_3 = R_p - \bar{r}_1 y_1 \tag{1.4.1}$$

$$y_2 + y_3 = 1 - y_1 \tag{1.4.2}$$

Multiplying (1.4.2) by $\bar{r}_3$ and subtracting it from (1.4.1) gives

$$(\bar{r}_2 - \bar{r}_3) y_2 = R_p - \bar{r}_1 y_1 - \bar{r}_3 + \bar{r}_3 y_1 \tag{1.4.3}$$

and hence

$$y_2 = \tilde{\alpha}_2 + \tilde{\beta}_2 y_1 \quad \text{where} \quad \tilde{\alpha}_2 = \frac{R_p - \bar{r}_3}{\bar{r}_2 - \bar{r}_3} \quad \text{and} \quad \tilde{\beta}_2 = \frac{\bar{r}_3 - \bar{r}_1}{\bar{r}_2 - \bar{r}_3}. \tag{1.4.4}$$

Moreover, (1.4.2) gives $y_3 = 1 - y_1 - y_2$ which simplifies to

$$y_3 = \tilde{\alpha}_3 + \tilde{\beta}_3 y_1 \quad \text{where} \quad \tilde{\alpha}_3 = 1 - \tilde{\alpha}_2 \quad \text{and} \quad \tilde{\beta}_3 = -(1 + \tilde{\beta}_2). \qquad (1.4.5)$$

If we write x in place of the unknown y_1 and also define

$$\tilde{\alpha} = \begin{pmatrix} 0 \\ \tilde{\alpha}_2 \\ \tilde{\alpha}_3 \end{pmatrix} \quad \text{and} \quad \tilde{\beta} = \begin{pmatrix} 1 \\ \tilde{\beta}_2 \\ \tilde{\beta}_3 \end{pmatrix}$$

then $y = \tilde{\alpha} + \tilde{\beta}x$. The risk V can now be expressed as a function of x, i.e.,

$$V = (\tilde{\alpha} + \tilde{\beta}x)^T Q(\tilde{\alpha} + \tilde{\beta}x) = \tilde{\alpha}^T Q\tilde{\alpha} + (2\tilde{\beta}^T Q\tilde{\alpha})x + (\tilde{\beta}^T Q\tilde{\beta})x^2. \qquad (1.4.6)$$

Hence

$$\frac{dV}{dx} = 2\tilde{\beta}^T Q\tilde{\alpha} + 2(\tilde{\beta}^T Q\tilde{\beta})x$$

and a stationary point of V occurs at

$$x = -\frac{\tilde{\beta}^T Q\tilde{\alpha}}{\tilde{\beta}^T Q\tilde{\beta}}. \qquad (1.4.7)$$

Exercises

1. Show that, for the asset data in Table 1.1, the mean returns $\bar{r}$ in (1.2.1) imply $\tilde{\alpha} \approx (0,\ 3.5,\ -2.5)^T$ and $\tilde{\beta} \approx (1,\ -6,\ 5)^T$ when the target return is $R_p = 1.2\%$. Hence show that the minimum risk strategy for an expected return of 1.2% is to spread the investment in the ratio 0.53 : 0.32 : 0.15 over assets 1, 2 and 3.

2. Find the minimum-risk portfolio that will give an expected return of 1.1% for the assets in Table 1.1.

3. For the data in Table 1.2, find the minimum-risk portfolio to achieve an expected return $R_p = 0.5\%$.

4. Suppose $y_1, \ldots y_4$ are invested fractions for a four-asset portfolio. Use (1.2.3) and (1.2.2) to obtain expressions – similar to (1.4.4) and (1.4.5) – for y_3 and y_4 in terms of y_1 and y_2.

5. Two- and three-asset minimum-risk solutions

It is useful to have computer implementations of the solution methods described in sections 3 and 4. The results given in this section have been obtained using a suite of software called SAMPO, which is written in Fortran90. It is *not* essential for the reader to understand or use this software (although it can be downloaded from an `ftp` site as described in the appendix). Solutions to **Minrisk0**, **Risk-Ret1** and **Minrisk1** can easily be coded in other languages or expressed as spreadsheet calculations. Results obtained with the `SAMPO` software will be quoted extensively this book in order to compare the performance of different optimization methods. However, we shall defer discussion of this until later chapters. For the moment we simply note that the results below are from a program `sample1` which is designed to read asset data (like that appearing in Tables 1.1 – 1.3) and then calculate solutions to **Minrisk0**, **Risk-Ret1** or **Minrisk1**.

The first problems we consider are based on data for the first two assets in Table 1.2. The expected returns turn out to be $\bar{r}_1 \approx 0.138$, $\bar{r}_2 \approx 0.004$ and the variance-covariance matrix is

$$Q \approx \begin{pmatrix} 0.0065 & -0.0039 \\ -0.0039 & 0.0029 \end{pmatrix}.$$

The solution to problem **Minrisk0** appears in Table 1.4.

y_1	y_2	R	V_{min}
0.397	0.603	0.057%	2.4×10^{-4}

Table 1.4. Solution of **Minrisk0** for first two assets in Table 1.2

Table 1.5 shows solutions to **Risk-Ret1** for the same data. These are obtained by minimizing (1.3.9) for various values of the weighting parameter ρ.

ρ	y_1	y_2	R	V
10	0.787	0.213	0.109%	2.85×10^{-4}
100	0.436	0.564	0.062%	2.7×10^{-4}
1000	0.401	0.599	0.058%	2.4×10^{-4}

Table 1.5. Solutions of **Risk-Ret1** for first two assets in Table 1.2

The values of y_1 and y_2 vary appreciably with ρ and tend towards the solution of **Minrisk0** as ρ gets larger. It can be shown (Exercise 3 below) that this will be the case for *any* asset data and is not just a feature of the present problem.

We now turn to problem **Minrisk1m**, still considering the first two assets in Table 1.2. Table 1.6 shows the solutions obtained by minimizing (1.3.15) for various values of ρ when $R_p = 0.1\%$. We observe that the choice of ρ is quite important. When ρ is small, the invested fractions which minimize (1.3.15) do not yield the target value for expected return; and it is only as ρ increases

ρ	y_1	y_2	R	V^*
0.01	0.56	0.44	0.079%	7.0×10^{-3}
0.1	0.688	0.312	0.096%	1.7×10^{-3}
1	0.713	0.287	0.1%	1.96×10^{-3}
10	0.716	0.284	0.1%	1.99×10^{-3}

Table 1.6. Solutions of **Minrisk1m** for first two assets in Table 1.2

that R tends to the required value $R_p = 0.1\%$. If we take $\rho > 10$ then – for this particular problem – there will not be any significant change in y_1 and y_2. (However, for different asset data, values of ρ very much greater than 10 might be needed before an acceptable solution to **Minrisk1** was obtained.)

We next consider a three-asset problem involving the data in Table 1.1. We can use the program `sample1` to solve **Minrisk1** by minimizing (1.4.6). Table 1.7 shows the results obtained for various values of target return R_p.

R_p	y_1	y_2	y_3	V^*
1.15%	0.295	0.229	0.475	4.4×10^{-3}
1.2%	0.530	0.321	0.149	1.3×10^{-3}
1.25%	0.764	0.413	-0.177	1.8×10^{-3}

Table 1.7. Solutions of **Minrisk1** for assets in Table 1.1

The first two rows of Table 1.7 show that a reduction in the value of R_p does not *necessarily* imply a reduced risk. If we decrease the target expected return from 1.2% to 1.15% we bias investment *away from* the better performing asset 1. This means that the positive co-variance σ_{23} in (1.2.7) will contribute more to V than the negative co-variances σ_{12}, σ_{13} and so the optimum value of risk is not so small. By comparing rows two and three of Table 1.7, however, we see that an increase in R_p from 1.2% to 1.25% does produce a corresponding increase in risk.

The negative value for y_3 in the third row of Table 1.7 is not an error. It indicates that, for a target return $R_p = 1.25\%$, the optimal investment strategy involves *short selling* – i.e., selling shares in asset 3 *even though the investor does not own them.* This can be done by borrowing the shares from a broker with the intention of returning them later. The strategy is only effective if the price of the shares falls because then the cost of buying replacement shares at a later date is less than the receipts from the sale of the borrowed ones.

We shall see in later sections how to formulate the problem so that only positive y_i are allowed – i.e., so that short-selling is not permitted. It is worth pointing out, however, that a minimum-risk solution is less likely to involve short-selling if we choose the target return R_p well within the range of the expected returns $\bar{r}_1,..,\bar{r}_n$ for the assets to be included in the portfolio.

By plotting the Table 1.7 values of y_1, y_2 and y_3 against R_p we can see that the invested fractions vary linearly with the target expected return. It is clear, however, that the relationship between minimum risk V^* and R_p is *not* linear. The reader can verify (Exercises 4 and 5 below) that when we plot V^* against R_p we obtain a parabolic curve. This is known as the *efficient frontier*.

Exercises (To be solved using `sample1` or other suitable software)
1. Find the minimum-risk portfolio for assets 2 and 3 in Table 1.2. Do the same calculation for assets 1 and 3.

2. Solve **Risk-Ret1** with $\rho = 1$ for assets 1 and 2 in Table 1.2. What is the smallest value of ρ that gives a solution without short-selling?

3. Show that the solution of **Risk-Ret1** tends to the solution of **Minrisk0** as $\rho \to \infty$.

4. For the data in Table 1.1, solve **Minrisk1** for $R_p = 1.175\%$ and $R_p = 1.225\%$. Combining these results with those in Table 1.7, plot the graph of V^* against R_p (the *efficient frontier*).

5. Solve **Minrisk1** for the data in Table 1.2 using several values of expected return in the range $0.004 \leq R_p \leq 0.138$. Hence plot the efficient frontier. For what range of R_p values does the minimum-risk portfolio avoid short-selling?

6. Calculate the mean returns and the variance-covariance matrix for the assets in Table 1.8.

	Return % for					
	Period 1	Period 2	Period 3	Period 4	Period 5	Period 6
Asset 1	0.4	0.5	0.55	0.61	0.65	0.7
Asset 2	0.4	0.41	0.42	0.43	0.44	0.45
Asset 3	0.7	0.65	0.61	0.55	0.5	0.4
Asset 4	0.4	0.41	0.42	0.43	0.44	0.45

Table 1.8. Rates of return on four assets

Using $\rho = 20$, determine the optimal portfolio given by solving **Risk-Ret1** for assets 1 and 2. How does this compare with the portfolio obtained from **Risk-Ret1** for assets 3 and 4? What do you notice about the mean returns for the four assets? Can you explain the differences between the optimal portfolios by looking at the Q-matrix?

6. A derivation of the minimum risk problem

So far we have merely stated that the function (1.2.11) is a measure of risk. Now we provide an explanation of the ideas behind problem **Minrisk0**.

Ideally, a "risk-free" portfolio would involve a combination of assets producing the same expected return R in all circumstances. In particular, looking at the history of m time periods, we would like $y_1,...,y_n$ to satisfy

$$r_{1j}y_1 + r_{2j}y_2 + .. + r_{nj}y_n = R$$

for $j = 1,...,m$. From the definition of R in (1.2.3), this is equivalent to

$$(r_{1j} - \bar{r}_1)y_1 + (r_{2j} - \bar{r}_2)y_2 + ... + (r_{nj} - \bar{r}_n)y_n = 0 \quad \text{for} \quad j = 1,...,m. \quad (1.6.1)$$

In general, however, this is an *overdetermined* set of equations (when $m > n$) which only has the solution $y_i = 0$. We cannot accept this because, by (1.2.14), we also require $e^T y = 1$. But if we cannot satisfy (1.6.1) exactly then the "next best thing" would be to make the residuals of the equations as small as possible. Let A denote the $m \times n$ matrix whose elements a_{ji} are defined by

$$a_{ji} = r_{ij} - \bar{r}_i. \quad (1.6.2)$$

Then (1.6.1) is equivalent to the system $Ay = 0$ and we want to choose y to minimize some norm $||Ay||$. It is convenient to use the two-norm

$$||Ay||_2 = \sqrt{(y^T A^T Ay)}.$$

By the rules of matrix multiplication the diagonal elements of $A^T A$ are

$$\sum_{j=1}^{m} a_{ji}^2 = \sum_{j=1}^{m} (r_{ij} - \bar{r}_i)^2, \quad \text{for } i = 1,...,n.$$

The off-diagonal (i,k)-th element of $A^T A$ is given by

$$\sum_{j=1}^{m} a_{ji} a_{jk} = \sum_{k=1}^{m} (r_{ij} - \bar{r}_i)(r_{kj} - \bar{r}_k).$$

Comparison with (1.2.4) and (1.2.5) shows that the i-th diagonal element of $A^T A$ is $m\sigma_i^2$ and that the (i,k)-th off-diagonal is $m\sigma_{ik}$. Hence $A^T A = mQ$ where Q is given by (1.2.12). Since the multiplying factor is irrelevant, we see that the problem of minimizing the risk function $y^T Qy$ is equivalent to minimizing the two-norm of the residuals of the system (1.6.1).

Exercises

1. Show that the variance-covariance matrix Q has the property that $y^T Qy \geq 0$ for all vectors y. Show also that $y^T Qy \neq 0$ for any $y \neq 0$ provided the columns of the matrix defined by (1.6.2) are linearly independent.

2. In the two-asset case, show that the function $V(x)$ defined by (1.3.3) has a non-negative second derivative. Show also that when $F(x)$ is defined by (1.3.9) or by (1.3.15) its second derivative is always positive.

7. Maximum return problems

All the problems considered so far have involved quadratic functions which can be minimized simply by solving a linear equation obtained by setting the first derivative to zero. However, not all portfolio optimization problems are quite so easy to deal with – as we shall now show. Suppose we want to fix an acceptable level of risk (as V_a, say) and then to *maximize* the expected return. This can be posed as the constrained minimization problem

Maxret1

$$\text{Minimize} \quad R = -\bar{r}^T y \tag{1.7.1}$$

$$\text{subject to} \quad V = y^T Qy = V_a \quad \text{and} \quad \sum_{i=1}^{n} y_i = 1. \tag{1.7.2}$$

To deal with (1.7.1), (1.7.2) we can proceed as in section 1.3 by introducing a composite function and setting up the modified problem

Maxret1m

$$\text{Minimize} \quad F = -\bar{r}^T y + \frac{\rho}{V_a^2}(y^T Qy - V_a)^2 \quad \text{subject to} \quad \sum_{i=1}^{n} y_i = 1 \tag{1.7.3}$$

where ρ is a weighting factor controlling the balance between return and risk. We can expect a solution of (1.7.3) to occur when $\bar{r}^T y$ is large and $y^T Qy$ is close to V_a. As before, we can use (1.2.2) to eliminate y_n and perform an unconstrained minimization of (1.7.3) with respect to $y_1, .., y_{n-1}$.

For two-asset problems we can write $y = \alpha + \beta x$ where $y_1 = x$ and α and β are given by (1.3.2). The function in (1.7.3) then becomes

$$F = -\bar{r}^T \alpha - (\bar{r}^T \beta)x + \frac{\rho}{V_a^2}[\alpha^T Q\alpha + 2(\beta^T Q\alpha)x + (\beta^T Q\beta)x^2 - V_a]^2. \tag{1.7.4}$$

If we consider the assets in Table 1.3 and substitute values for $\bar{r}$ and Q from (1.3.6) then (1.7.4) becomes

$$-0.06667x - 1.1167 + \frac{\rho}{V_a^2}(0.1256x^2 - 0.1589x + 0.05139 - V_a)^2. \quad (1.7.5)$$

This is a *quartic* function and so its first derivative will be a cubic rather than a linear expression; hence the calculation of stationary points is no longer trivial.

A similar approach to the maximum-return problem could be based on using the absolute value (rather than the square) of the difference between the actual portfolio risk and the acceptable level V_a. This would lead to solving

$$\text{Minimize} \quad F = -\bar{r}^T y + \frac{\rho}{V_a}|y^T Q y - V_a| \quad \text{subject to} \quad \sum_{i=1}^{n} y_i = 1. \quad (1.7.6)$$

Reducing (1.7.6) to a one-variable problem leads to the minimization of

$$F = -\bar{r}^T\alpha - (\bar{r}^T\beta)x + \frac{\rho}{V_a}|\alpha^T Q\alpha + 2(\beta^T Q\alpha)x + (\beta^T Q\beta)x^2 - V_a|. \quad (1.7.7)$$

If we substitute data from Table 1.3 into (1.7.7) we get

$$F = -0.06667x - 1.1167 + \frac{\rho}{V_a}|0.1256x^2 - 0.1589x + 0.05139 - V_a|. \quad (1.7.8)$$

This function is *non-differentiable* at all values of x for which the expression inside the absolute value signs is zero – i.e. whenever

$$0.1256x^2 - 0.1589x + 0.05139 = V_a.$$

If we draw a graph of (1.7.8), say for $\rho = 10, V_a = 0.002$ we can see why functions like (1.7.8) are called *non-smooth*.

Non-quadratic and non-smooth functions like (1.7.5) and (1.7.8) cannot be minimized simply by forming and solving a linear equation given by setting the first derivative to zero. Instead we need to use iterative methods of the kind described in the next chapter.

Exercises

1. Calculate the function (1.7.4) using data for the first two assets in Table 1.2. Plot a graph of this function for $V_a = 0.0009$ and $\rho = 10$ and hence estimate its minimum.

2. Using the data for question 1, plot the function (1.7.7) for $V_a = 0.0009$ and $\rho = 10$ and estimate its minimum.

Chapter 2

ONE-VARIABLE OPTIMIZATION

1. Optimality conditions

We begin with a formal statement of the conditions which hold at a minimum of a one-variable differentiable function. We have already made use of these conditions in the previous chapter.

Definition Suppose that $F(x)$ is a continuously differentiable function of the scalar variable x, and that, when $x = x^*$,

$$\frac{dF}{dx} = 0 \quad \text{and} \quad \frac{d^2F}{dx^2} > 0. \qquad (2.1.1)$$

The function $F(x)$ is then said to have a *local minimum* at x^*.

Conditions (2.1.1) imply that $F(x^*)$ is the smallest value of F in some region near x^*. It may also be true that $F(x^*) \leq F(x)$ for *all* x but condition (2.1.1) does not guarantee this.

Definition If conditions (2.1.1) hold at $x = x^*$ and if $F(x^*) \leq F(x)$ for *all* x then x^* is said to be the *global minimum*.

In practice it is usually hard to establish that x^* is a global minimum and so we shall chiefly be concerned with methods of finding local minima.

Conditions (2.1.1) are called *optimality conditions*. For simple problems they can be used directly to find a minimum, as in section 3 and 1.3 of chapter 1. As another example, consider the function

$$F(x) = x^3 - 3x^2. \qquad (2.1.2)$$

Here, and in what follows, we shall sometimes use the notation

$$F'(x) = \frac{dF}{dx} \quad \text{and} \quad F''(x) = \frac{d^2F}{dx^2}.$$

Thus, for (2.1.2), $F'(x) = 3x^2 - 6x$ and so $F'(x) = 0$ when $x = 0$ and $x = 2$. These values represent two *stationary points* of $F(x)$; and to determine which is a minimum we must consider $F''(x) = 6x - 6$. We see that F has a minimum at $x = 2$ because $F''(2) > 0$. However, $F''(0)$ is negative and so $F(x)$ has a local maximum at $x = 0$.

We can only use this analytical approach to solve minimization problems when it is easy to form and solve the equation $F'(x) = 0$. This may not be the case for functions $F(x)$ which occur in practical problems – for instance **Maxret1m**. Therefore we shall usually resort to iterative techniques.

Some iterative methods are called *direct search* techniques and are based on simple comparison of function values at trial points. Others are known as *gradient methods*. These use derivatives of the objective function and can be viewed as iterative algorithms for solving the nonlinear equation $F'(x) = 0$. Gradient methods tend to converge faster than direct search methods. They also have the advantage that they permit an obvious convergence test – namely stopping the iterations when the gradient is near zero. Gradient methods are not always suitable, however – i.e. when $F(x)$ has discontinuous derivatives, as on a piecewise linear function. We have already seen an example of this in function (1.7.8).

Exercises

1. Show that, if the optimality conditions (2.1.1) hold, $F(x^* + h) > F(x^*)$ for h sufficiently small.

2. Find the stationary points of $F(x) = 4\cos x^2 - \sin x^2 - 3$.

2. The bisection method

A simple (but inefficient) way of estimating the least value of $F(x)$ in a range $a \leq x \leq b$ would be calculate the function at many points in $[a, b]$ and then pick the one with lowest value. The *bisection method* uses a more systematic approach to the evaluation of F in $[a, b]$. Each iteration uses a comparison between function values at five points to reduce by half the size of a bracket containing the minimum. It can be shown that this approach will locate a minimum if it is applied to a function $F(x)$ that is *unimodal* – i.e., one that has only one minimum in the range $[a, b]$.

A statement of the algorithm is given below.

Bisection Method for minimizing $F(x)$ on the range $[a,b]$

Set $x_a = a$, $x_b = b$ and $x_m = \frac{1}{2}(a+b)$.
Calculate $F_a = F(x_a)$, $F_b = F(x_b)$, $F_m = F(x_m)$
Repeat
set $x_l = \frac{1}{2}(x_a + x_m)$, $x_r = \frac{1}{2}(x_m + x_b)$
calculate $F_l = F(x_l)$ and $F_r = F(x_r)$
let $F_{min} = \min\{F_a, F_b, F_m, F_l, F_r\}$
if $F_{min} = F_a$ or F_l then set $x_b = x_m$, $x_m = x_l$, $F_b = F_m$, $F_m = F_l$
else if $F_{min} = F_m$ then set $x_a = x_l$, $x_b = x_r$, $F_a = F_l$, $F_b = F_r$
else if $F_{min} = F_r$ or F_b then set $x_a = x_m$, $x_m = x_r$, $F_a = F_m$, $F_m = F_r$
until $|x_b - x_a|$ is sufficiently small

Proposition If $F(x)$ is unimodal for $a \le x \le b$ with a minimum at x^* then the number of iterations taken by the bisection method to locate x^* within a bracket of width less than 10^{-s} is K, where K is the smallest integer which exceeds

$$\frac{log_{10}(b-a)+s}{log_{10}(2)}. \tag{2.2.1}$$

Proof The size of the bracket containing the solution is halved on each iteration. Hence, after k iterations the width of the bracket is $2^{-k}(b-a)$. To find the value of k which gives

$$2^{-k}(b-a) = 10^{-s}$$

we take logs of both sides and get

$$log_{10}(b-a) - k log_{10}(2) = -s$$

and so the width of the bracket is less than 10^{-s} once k exceeds (2.2.1).

We can show how the bisection method works by applying it to the problem

$$\text{Minimize } F(x) = x^3 - 3x^2 \text{ for } 0 \le x \le 3.$$

Initially $x_a = 0$, $x_b = 3$, $x_m = 1.5$, and the first iteration adds the points $x_l = 0.75$ and $x_r = 2.25$. We then find

$$F_a = 0;\ F_l = -1.266;\ F_m = -3.375;\ F_r = -3.797;\ F_b = 0.$$

The least function value is at $x_r = 2.25$ and hence the search range for the next iteration is $[x_m, x_b] = [1.5, 3.0]$. After re-labelling the points and computing new values x_l, x_r we get

$$x_a = 1.5;\ x_l = 1.875;\ x_m = 2.25;\ x_r = 2.625;\ x_b = 3$$

and $F_a = -3.375;\ F_l = -3.955;\ F_m = -3.797;\ F_r = -2.584;\ F_b = 0.$

Now the least function value is at x_l and the new range is $[x_a, x_m] = [1.5,\ 2.25]$. Re-labelling and adding the new x_l and x_r we get

$$x_a = 1.5;\ x_l = 1.6875;\ x_m = 1.875;\ x_r = 2.0625;\ x_b = 2.25$$

and $F_a = -3.375;\ F_l = -3.737;\ F_m = -3.955;\ F_r = -3.988;\ F_b = -3.797.$

These values imply the minimum lies in $[x_l, x_r] = [1.875, 2.25]$. After a few more steps we have an acceptable approximation to the true solution at $x = 2$.

The application of the bisection method to a maximum-return problem is illustrated below in section 6 of this chapter.

Finding a bracket for a minimum

We now give a systematic way of finding a range $a < x < b$ which contains a minimum of $F(x)$. This method uses the slope F' to determine whether the search for a minimum should be to the left or the right of an initial point x_0. If $F'(x_0)$ is positive then lower values of the function will be found for $x < x_0$, while $F'(x_0) < 0$ implies that lower values of F occur when $x > x_0$. The algorithm merely takes larger and larger steps in a "downhill" direction until the function starts to increase, indicating that a minimum has been bracketed.

Finding a and b to bracket a local minimum of $F(x)$

Choose an initial point x_0 and a step size $\alpha(> 0)$
Set $\delta = -\alpha \times sign(F'(x_0))$
Repeat for $k = 0, 1, 2, ...$
$x_{k+1} = x_k + \delta, \quad \delta = 2\delta$
until $F(x_{k+1}) > F(x - k)$
if $k = 0$ then set $a = x_0$ and $b = x_1$
if $k > 0$ then set $a = x_{k-1}$ and $b = x_{k+1}$

Exercises

1. Apply the bisection method to $F(x) = e^x - 2x$ in the interval $0 \leq x \leq 1$.

2. Do two iterations of the bisection method for the function $F(x) = x^3 + x^2 - x$ in the range $0 \leq x \leq 1$. How close is the best point found to the exact minimum of F? What happens if we apply the bisection method in the range $-2 \leq x \leq 0$?

3. Use the bracketing technique with $x_0 = 1$ and $\alpha = 0.1$ to find a bracket for a minimum of $F(x) = e^x - 2x$.

4. Apply the bracketing algorithm to the function (1.7.5) with $V_a = 0.00123$ and $\rho = 10$ and using $x_0 = 0$ and $\alpha = 0.25$ to determine a starting range for

the bisection method. How does the result compare with the bracket obtained when $x_0 = 1$ and $\alpha = 0.2$?

3. The secant method

We now consider an iterative method for solving $F'(x) = 0$. This will find a local minimum of $F(x)$ *provided we use it in a region where the second derivative $F''(x)$ remains positive.* The approach is based on linear interpolation. If we have evaluated F' at two points $x = x_1$ and $x = x_2$ then the calculation

$$x = x_1 - \frac{F'(x_1)}{F'(x_2) - F'(x_1)}(x_2 - x_1) \tag{2.3.1}$$

gives an estimate of the point where F' vanishes.

We can show that formula (2.3.1) is exact if F is a quadratic function. Consider $F(x) = x^2 - 3x - 1$ for which $F'(x) = 2x - 3$ and suppose we use $x_1 = 0$ and $x_2 = 2$. Then (2.3.1) gives

$$x = 0 - \frac{F'(0)}{F'(2) - F'(0)} \times 2 = 0 - \frac{-3}{4} \times 2 = 1.5.$$

Hence (2.3.1) has yielded the minimum of $F(x)$. However, when F is not quadratic, we have to apply the interpolation formula iteratively. The algorithm below shows how this might be done.

Secant method for solving $F'(x) = 0$

Choose x_0, x_1 as two estimates of the minimum of $F(x)$
Repeat for $k = 0, 1, 2...$

$$x_{k+2} = x_k - \frac{F'(x_k)}{F'(x_{k+1}) - F'(x_k)}(x_{k+1} - x_k)$$

until $|F'(x_{k+2})|$ is sufficiently small.

This algorithm makes repeated use of formula (2.3.1) based upon the two most recently calculated points. In fact, this may not be the most efficient way to proceed. When $k > 1$, we would normally calculate x_{k+2} using x_{k+1} together with *either* x_k *or* x_{k-1} according to one of a number of possible strategies:
(a) Choose whichever of x_k and x_{k-1} gives the smaller value of $|F'|$.
(b) Choose whichever of x_k and x_{k-1} gives F' with opposite sign to $F'(x_{k+1})$
(c) Choose whichever of x_k and x_{k-1} gives the smaller value of F.
Strategies *(a)* and *(c)* are based on using points which seem closer to the minimum; strategy *(b)* reflects the fact that linear interpolation is more reliable than

linear extrapolation (which occurs when $F'(x_{k+1})$ and $F'(x_k)$ have the same sign). Strategy *(b)*, however, can only be employed if we have chosen our initial x_0 and x_1 so that $F'(x_0)$ and $F'(x_1)$ have opposite signs.

We can demonstrate the secant method with strategy *(a)* on $F(x) = x^3 - 3x^2$ for which $F'(x) = 3x^2 - 6x$. If $x_0 = 1.5$ and $x_1 = 3$ then $F'(x_0) = -2.25$ and $F'(x_1) = 9$. Hence the first iteration gives

$$x_2 = 1.5 - \frac{-2.25}{11.25} \times 1.5 = 1.8$$

and so $F'(x_2) = -1.08$. For the next iteration we re-assign $x_1 = 1.5$, since $|F'(1.5)| < |F'(3)|$, and so we get

$$x_3 = 1.5 - \frac{-2.25}{-1.17} \times 0.3 \approx 2.077.$$

The iterations appear to be converging towards the correct solution at $x = 2$, and the reader can perform further calculations to confirm this.

Exercises

1. Apply the secant method to $F(x) = e^x - 2x$ in the range $0 \leq x \leq 1$.

2. Show that (2.3.1) will give $F'(x) = 0$ when applied to *any* quadratic function

$$F(x) = ax^2 + bx + c.$$

3. Use the secant method with strategy *(b)* on $F(x) = x^3 - 3x^2$ with $x_0 = 1.5$ and $x_1 = 3$. What happens if the starting values are $x_0 = 0.5$ and $x_1 = 1.5$?

4. The Newton method

This method seeks the minimum of $F(x)$ using both first *and* second derivatives. In its simplest form it is described as follows.

Newton method for minimizing $F(x)$

Choose x_0 as an estimate of the minimum of $F(x)$
Repeat for $k = 0, 1, 2...$

$$x_{k+1} = x_k - \frac{F'(x_k)}{F''(x_k)} \tag{2.4.1}$$

until $|F'(x_{k+1})|$ is sufficiently small.

This algorithm is derived by expanding $F(x)$ as a Taylor series about x_k

$$F(x_k + h) = F(x_k) + hF'(x_k) + \frac{h^2}{2}F''(x_k) + O(h^3). \tag{2.4.2}$$

Differentiation with respect to h gives a Taylor series for $F'(x)$

$$F'(x_k + h) = F'(x_k) + hF''(x_k) + O(h^2). \quad (2.4.3)$$

If we assume that h is small enough for the $O(h^2)$ term in (2.4.3) to be neglected then it follows that the step $h = -F'(x_k)/F''(x_k)$ will give $F'(x_k + h) \approx 0$.

As an illustration, we apply the Newton method to $F(x) = x^3 - 3x^2$ for which $F'(x) = 3x^2 - 6x$ and $F''(x) = 6x - 6$. At the initial guess $x_0 = 3$, $F' = 9$ and $F'' = 12$ and so the next iterate is given by

$$x_1 = 3 - \frac{9}{12} = 2.25.$$

Iteration two uses $F'(2.25) = 1.6875$ and $F''(2.25) = 7.5$ to give

$$x_2 = 2.25 - \frac{1.6875}{7.5} = 2.025.$$

After one more iteration $x_3 \approx 2.0003$ and so Newton's method is converging to the solution $x^* = 2$ more quickly than either bisection or the secant method.

Convergence of the Newton method

Because the Newton iteration is important in the development of optimization methods we study its convergence more formally. We define

$$e_k = x^* - x_k \quad (2.4.4)$$

as the error in the approximate minimum after k iterations.

Proposition Suppose that the Newton iteration (2.4.1) converges to x^*, a local minimum of $F(x)$, and that $F''(x^*) = m > 0$. Suppose also that there is some neighbourhood, N, of x^* in which the third derivatives of F are bounded, so that, for some $M > 0$,

$$M \geq F'''(x) \geq -M \text{ for all } x \in N. \quad (2.4.5)$$

If e_k is defined by (2.4.4) then there exists an integer K such that, for all $k > K$,

$$\frac{e_k^2 M}{m} > e_{k+1} > -\frac{e_k^2 M}{m}. \quad (2.4.6)$$

Proof Since the iterates x_k converge to x^* there exists an integer K such that

$$x_k \in N \text{ and } |e_k| < \frac{m}{2M} \quad \text{for } k > K.$$

Then the bounds (2.4.5) on F''' imply

$$m + M|e_k| > F''(x_k) > m - M|e_k|.$$

Combining this with the bound on $|e_k|$, we get

$$F''(x_k) > \frac{m}{2}. \qquad (2.4.7)$$

Now, by the mean value form of Taylor's theorem ,

$$F'(x^*) = F'(x_k) + e_k F''(x_k) + \frac{1}{2} e_k^2 F'''(\xi),$$

for some ξ between x^* and x_k; and since $F'(x^*) = 0$ we deduce

$$F'(x_k) = -e_k F''(x_k) + \frac{1}{2} e_k^2 F'''(\xi).$$

The next estimate of the minimum is $x_{k+1} = x_k - \delta x_k$ where

$$\delta x_k = \frac{F'(x_k)}{F''(x_k)} = -e_k + \frac{e_k^2 F'''(\xi)}{2F''(x_k)}.$$

Hence the error after $k+1$ iterations is

$$e_{k+1} = x^* - x_{k+1} = e_k + \delta x_k = \frac{e_k^2 F'''(\xi)}{2F''(x_k)}.$$

Thus (2.4.6) follows, using (2.4.5) and (2.4.7).

This result shows that, when x_k is near to x^*, the error e_{k+1} is proportional to e_k^2 and so the Newton method ultimately approaches the minimum very rapidly.

Definition If, for some constant C, the errors e_k, e_{k+1} on successive steps of an iterative method satisfy

$$|e_{k+1}| \le C e_k^2 \quad \text{as } k \to \infty$$

then the iteration is said to have a *quadratic* rate of ultimate convergence.

Implementation of the Newton method

The convergence result in section 4 depends on certain assumptions about higher derivatives and this should warn us that the basic Newton iteration may not always converge. For instance, if the iterations reach a point where $F''(x)$ is zero the calculation will break down. It is not only this extreme case which can cause difficulties, however, as the following examples show.

Consider the function $F(x) = x^3 - 3x^2$, and suppose the Newton iteration is started from $x_0 = 1.1$. Since $F'(x) = 3x^2 - 6x$ and $F''(x) = 6x - 6$, we get

$$x_1 = 1.1 - \frac{(-2.97)}{0.6} = 6.05.$$

However, the minimum of $x^3 - 3x^2$ is at $x = 2$; and hence the method has overshot the minimum and given x_1 *further away* from the solution than x_0.

Suppose now that the Newton iteration is applied to $x^3 - 3x^2$ starting from $x_0 = 0.9$. The new estimate of the minimum turns out to be

$$x_1 = 0.9 - \frac{(-1.89)}{(-0.6)} = -2.25,$$

and the direction of the Newton step is away from the minimum. The iteration is being attracted to the maximum of $F(x)$ at $x = 0$. (Since the Newton method solves $F'(x) = 0$, this is not an unreasonable outcome.)

These two examples show that convergence of the basic Newton iteration depends upon the behaviour of $F''(x)$ and a practical algorithm should include safeguards against divergence. We should only use (2.4.1) if $F''(x)$ is strictly positive; and even then we should also check that the new point produced by the Newton formula is "better" than the one it replaces. These ideas are included in the following algorithm which applies the Newton method within a bracket $[a, b]$ such as can be found by the algorithm in section 2.

Safeguarded Newton method for minimizing $F(x)$ in $[a, b]$

Make a guess x_0 $(a < x_0 < b)$ for the minimum of $F(x)$
Repeat for $k = 0, 1, 2, ..$
 if $F''(x_k) > 0$ then $\delta x = -F'(x_k)/F''(x_k)$
 else $\delta x = -F'(x_k)$
 if $\delta x < 0$ then $\alpha = \min(1, (a - x_k)/\delta x)$
 if $\delta x > 0$ then $\alpha = \min(1, (b - x_k)/\delta x)$
 Repeat for $j = 0, 1, ...$
  $\alpha = 0.5^j \alpha$
 until $F(x_k + \alpha\delta x) < F(x_k)$
 Set $x_{k+1} = x_k + \alpha\delta x$
until $|F'(x_{k+1})|$ is sufficiently small.

As well as giving an alternative choice of δx when $F'' \leq 0$, the safeguarded Newton algorithm includes a step-size, α. This is chosen first to prevent the correction steps from going outside the bracket $[a, b]$ and then, by repeated halving, to ensure that each new point has a lower value of F than the previous one. The algorithm always tries the full step ($\alpha = 1$) first and hence it can have the same fast ultimate convergence as the basic Newton method.

We can show the working of the safeguarded Newton algorithm on the function $F(x) = x^3 - 3x^2$ in the range [1,4] with $x_0 = 1.1$. Since

$$F(1.1) = -2.299, \; F'(1.1) = -2.97 \text{ and } F''(1.1) = 0.6$$

the first iteration gives $\delta x = 4.95$. The full step, $\alpha = 1$, gives $x_k + \alpha\delta x = 6.05$ which is outside the range we are considering and so we must re-set

$$\alpha = \frac{(4 - 1.1)}{4.95} \approx 0.5859.$$

However, $F(4) = 16 > F(1.1)$ and α is reduced again (to about 0.293) so that

$$x_k + \alpha\delta x = 1.1 + 0.293 \times 4.95 \approx 2.45.$$

Now $F(2.45) \approx -3.301$ which is less than $F(1.1)$. Therefore the inner loop of the algorithm is complete and the next iteration can begin.

Under certain assumptions, we can show that the inner loop of the safeguarded Newton algorithm will always terminate and hence that the safeguarded Newton method will converge.

Exercises

1. Use Newton's method to estimate the minimum of $e^x - 2x$ in $0 \le x \le 1$. Compare the rate of convergence with that of the bisection method.

2. Show that, for any starting guess, the basic Newton algorithm converges in one step when applied to a quadratic function.

3. Do one iteration of the basic Newton method on the function $F(x) = x^3 - 3x^2$ starting from each of the following initial guesses:

$$x_0 = 2.1, \quad x_0 = 1, \quad x_0 = -1.$$

Explain what happens in each case.

4. Do two iterations of the safeguarded Newton method applied to the function $x^3 - 3x^2$ and starting from $x_0 = 0.9$.

5. Methods using quadratic or cubic interpolation

Each iteration of Newton's method generates x_{k+1} as a stationary point of the interpolating quadratic function defined by the values of $F(x_k)$, $F'(x_k)$ and $F''(x_k)$. In a similar way, a direct-search iterative approach can be based on locating the minimum of the quadratic defined by values of F at three points x_k, x_{k-1}, x_{k-2}; and a gradient approach could minimize the local quadratic approximation given by, say, $F(x_{k-1})$, $F'(x_{k-1})$ and $F(x_k)$. If a quadratically predicted minimum x_{k+1} is found to be "close enough" to x^* (e.g. because $F'(x_{k+1}) \approx 0$) then the iteration terminates; otherwise x_{k+1} is used instead of one of the current points to generate a new quadratic model and hence predict a new minimum.

As with the Newton method, the practical implementation of this basic idea requires certain safeguards, mostly for dealing with cases where the interpolated quadratic has negative curvature and therefore does not have a minimum. The bracketing algorithm given earlier may prove useful in locating a group of points which implies a suitable quadratic model.

A similar approach is based on repeated location of the minimum of a *cubic* polynomial fitted either to values of F at four points or to values of F and F' at two points. This method can give faster convergence, but it also requires fall-back options to avoid the search being attracted to a maximum rather than a minimum of the interpolating polynomial.

Exercises

1. Suppose that $F(x)$ is a quadratic function and that, for any two points x_a, x_b, the ratio D is defined by

$$D = \frac{F(x_b) - F(x_a)}{(x_b - x_a)F'(x_a)}.$$

Show that $D = 0.5$ when x_b is the minimum of $F(x)$. What is the expression for D if $F(x)$ is a cubic function?

2. Explain why the secant method can be viewed as being equivalent to quadratic interpolation for the function $F(x)$.

3. Design an algorithm for minimizing $F(x)$ by quadratic interpolation based on function values only.

6. Solving maximum-return problems

The `SAMPO` software was mentioned briefly in section 5 of chapter 1. The program `sample2` is designed to read asset data like that in Table 1.3 and then to solve problem **Maxret1m** using either the bisection, secant or Newton method. In this section we shall quote some results from `sample2` in order to compare the performance of the three techniques. The reader should be able to obtain similar comparative results using other implementations of these one-variable optimization algorithms. (It should be understood, however, that two different versions of the same iterative optimization method may not give *exactly* the same sequence of iterates even though they eventually converge to the same solution. Minor discrepancies can arise because of rounding errors in computer arithmetic when the same calculation is expressed in two different ways or when two computer systems work to different precision.)

We recall that the maximum-return problem **Maxret1m** involves minimizing the function (1.7.4). In particular, using data in Table 1.3, (1.7.4) becomes

$$-0.06667x - 1.1167 + \frac{\rho}{V_a^2}(0.1256x^2 - 0.1589x + 0.05139 - V_a)^2 \qquad (2.6.1)$$

where ρ is a weighting parameter and V_a denotes an acceptable value for risk. In order to choose a sensible value for V_a it can be helpful, first of all, to solve **Minrisk0** to obtain the *least possible* value of risk, V_{min}. For the data in Table 1.3 we find that $V_{min} \approx 0.00112$. Suppose now that we seek the maximum return when the acceptable risk is $V_a = 1.5V_{min} \approx 0.00168$. Table 2.1 shows results obtained when we use the bisection, secant and Newton methods to minimize (2.6.1) with $\rho = 1$.

Method	itns	y_1	y_2	V	R
Bisection	14	0.7	0.3	0.00169	1.233%
Secant	10	0.7	0.3	0.00169	1.233%
Newton	7	0.7	0.3	0.00169	1.233%

Table 2.1. Solutions of **Maxret1m** for assets in Table 1.3

We see that all three minimization methods find the same solution but that they use different numbers of iterations.

When drawing conclusions from a comparison like that in Table 2.1 it is important to ensure that all the methods have used the same (or similar) initial guessed solutions and convergence criteria. In the above results, both bisection and the secant method were given the starting values $x = 0$ and $x = 1$. The Newton method was started from $x = 1$ (it only needs one initial point). The bisection iterations terminate when the bracket containing the optimum has been reduced to a width less than 10^{-4}. Convergence of the secant and Newton methods occurs when the gradient $|F'(x)| < 10^{-5}$. Therefore it seems reasonable to conclude that the Newton method is indeed more efficient than the secant method which in turn is better than bisection.

We now consider another problem to see if similar behaviour occurs. This time we use data for the first two assets in Table 1.2 and we consider the minimization of (1.7.4) with $\rho = 10$ and $V_a = 0.00072$. Table 2.2 summarises results obtained with `sample2` using the same starting guessed values and convergence criteria as for Table 2.1 together with an additional result for Newton's method when the initial solution estimate is $x_0 = 1$.

Method	itns	y_1	y_2	V	R
Bisection	14	0.23	0.77	0.00072	0.035%
Secant	6	0.23	0.77	0.00072	0.035%
Newton ($x_0 = 0$)	5	0.23	0.77	0.00072	0.035%
Newton ($x_0 = 1$)	7	0.564	0.436	0.00072	0.08%

Table 2.2. Solutions of **Maxret1m** for first two assets in Table 1.2

The first three rows of Table 2.2 again show the secant and Newton methods outperforming bisection by finding the same solution in fewer iterations. In fact the number of bisection steps depends *only* on the size of the starting bracket and the convergence criterion while the iteration count for the secant and Newton methods can vary from one problem to another.

The last row of Table 2.2 shows that a different solution is obtained if Newton's method is started from $x_0 = 1$ instead of $x_0 = 0$. This alternative solution is actually better than the one in the first three rows of the table because the return R is larger. However, both the solutions are valid local minima of (1.7.4) and we could say, therefore, that the bisection and secant methods have been "unlucky" in converging to the inferior one. As mentioned in section 1, most of the methods covered in this book will terminate when the iterations reach a *local* optimum. If there are several minima, it is partly a matter of chance which one is found, although the one "nearest" to the starting guess is probably the strongest contender.

Exercises (To be solved using `sample2` or other suitable software.)

1. Using the data for the first two assets in Table 1.2, determine the coefficients of the function (1.7.4). Hence find the maximum return for an acceptable risk $V_a = 0.0005$ by using the bisection method to minimize (1.7.4). How does your solution change for different values of ρ in the range $0.1 \le \rho \le 10$?

2. Solve the maximum-return problem in question 1 but using the bisection method to minimize the non-smooth function (1.7.7). Explain why the results differ from those in question 1.

3. Using data for the first two assets in Table 1.2, form the function (1.7.4) and minimize it by the secant method when $V_a = 0.002$ and $\rho = 10$. Use starting guesses 0 and 1 for x. Can you explain why a different solution is obtained

when the starting guesses for x are 0.5 and 1? (It may help to sketch the function being minimized.)

4. Minimize (2.6.1) by Newton's method using the initial guess $x_0 = 0.5$ and explain why the solution is different from the ones quoted in Table 2.1. Find starting ranges for the bisection and secant methods from which they too will converge to this alternative local minimum.

5. Plot the graph of (2.6.1) for $0.55 \leq x \leq 0.75$, when $V_a = 0.00123$ and $\rho = 10$ and observe the two local minima. Also plot the graph when $V_a = 0.001$. Can you explain why there is a unique minimum in this case? What is the smallest value of V_a for which two minima occur?

6. Plot the graph of (1.7.8) with $\rho = 1$ in the range $0.55 \leq x \leq 0.75$ first for $V_a = 0.001$ and then for $V_a = 0.00123$. Does the function have a continuous first derivative in both cases? Explain your answer.

7. Using any suitable optimization method, minimize (2.6.1) and hence solve the maximum-return problem for the data in Table 1.3 for values of V_a in the range $0.0011 \leq V_a \leq 0.0033$. Plot the resulting values of y_1 and y_2 against V_a. Do these show a linear relationship? How does the graph of maximum-return R against V_a compare with the efficient frontier you would obtain by solving **Minrisk1m** for the data in Table 1.3?

8. As an alternative to the composite function in **Risk-Ret1**, another function whose minimum value will give a low value of risk coupled with a high value for expected return is

$$F = \frac{y^T Q y}{\bar{r}^T y}. \tag{2.6.2}$$

For a two-asset problem, use similar ideas to those in section 1.3 of chapter 1 and express F as a function of invested fraction y_1 only. Use the bisection method to minimize this one-variable form of (2.6.2) for the data in Table 1.3. Also obtain an expression for dF/dy_1 and use the secant method to estimate the minimum of F for the same set of data.

Chapter 3

OPTIMAL PORTFOLIOS WITH N ASSETS

1. Introduction

So far we have only dealt with portfolios involving two or three assets which lead to optimization problems in one variable. For a larger portfolios with n assets we can generalise the ideas in sections 3 and 7 of chapter 1 in which we eliminate one of the invested fractions and obtain an optimization problem in $n-1$ variables.

Let us recall the notation and terminology for an n-asset problem involving a history of returns over m time periods. If we have computed the mean returns $\bar{r}_i$ for each asset as well as the variances σ_{ii} and the covariances σ_{ij} then the expected return on a portfolio defined by invested fractions $y_1, \ldots y_n$ is given by

$$R = \bar{r}^T y \quad \text{where} \quad \bar{r} = \begin{pmatrix} \bar{r}_1 \\ \bar{r}_2 \\ \cdot\cdot \\ \bar{r}_n \end{pmatrix} \quad \text{and} \quad y = \begin{pmatrix} y_1 \\ y_2 \\ \cdot\cdot \\ y_n \end{pmatrix}. \tag{3.1.1}$$

The expression for risk can be written as

$$V = y^T Q y \tag{3.1.2}$$

where

$$Q = \begin{pmatrix} \sigma_{11} & \sigma_{12} & \ldots & \sigma_{in} \\ \sigma_{12} & \sigma_{22} & \ldots & \sigma_{2n} \\ \ldots & \ldots & \ldots & \ldots \\ \ldots & \sigma_{ij} & \ldots & \ldots \\ \ldots & \ldots & \ldots & \ldots \\ \sigma_{1n} & \sigma_{2n} & \ldots & \sigma_{nn} \end{pmatrix}. \tag{3.1.3}$$

We also recall the normalisation condition $\sum_{i=1}^{n} y_i = 1$, which can be written

$$S = e^T y = 1, \quad \text{where } e = \begin{pmatrix} 1 \\ 1 \\ .. \\ 1 \end{pmatrix}. \tag{3.1.4}$$

We can use (3.1.4) to eliminate y_n and hence express return and risk in terms of $y_1, ..., y_{n-1}$ only. Generalising the approach used in section 3 of chapter 1, we let x be the column vector with $x_i = y_i, (i = 1, ..., n-1)$. We also define an $(n-1)$-vector, α, and an $n \times (n-1)$ matrix, B, as

$$\alpha = \begin{pmatrix} 0 \\ 0 \\ .. \\ 0 \\ 1 \end{pmatrix} \quad \text{and} \quad B = \begin{pmatrix} 1 & 0 & ... & 0 \\ 0 & 1 & ... & 0 \\ .. & .. & .. & .. \\ 0 & 0 & ... & 1 \\ -1 & -1 & ... & -1 \end{pmatrix}. \tag{3.1.5}$$

These are the n-asset extensions of (1.3.2) and it is easy to see that they lead to

$$y = \alpha + Bx. \tag{3.1.6}$$

We can now formulate **Minrisk0**, **Minrisk1** and **Maxret1** for the n-asset case.

2. The basic minimum-risk problem

If we use the transformation (3.1.5) and (3.1.6) then problem **Minrisk0** is equivalent to

$$\text{Minimize} \quad V = (\alpha + Bx)^T Q(\alpha + Bx) \tag{3.2.1}$$

which is an unconstrained optimization problem in $n-1$ variables. If x^* solves (3.2.1) then the minimum-risk portfolio is

$$y_i^* = x_i^*, \;\; i = 1, .., n-1; \;\; y_n^* = 1 - \sum_{i=1}^{n-1} x_i^*. \tag{3.2.2}$$

Many methods for minimizing a function of several variables involve the vector of first partial derivatives. A common notation is

$$\nabla V = (\frac{\partial V}{\partial x_1}, \frac{\partial V}{\partial x_2} ...)^T$$

where ∇V is called the *gradient* of V. (An alternative notation for ∇V is V_x.) The reader can verify, by fairly simple algebra, that for V in (3.2.1)

$$\nabla V = 2B^T Q\alpha + 2B^T QBx. \tag{3.2.3}$$

We may also need to consider second partial derivatives of functions appearing in optimization problems. Thus, for problem (3.2.1), we can form the *Hessian matrix* $\nabla^2 V$ (sometimes written V_{xx}) whose (i,j)-th element is $\partial^2 V/\partial x_i \partial x_j$. It is straightforward to show that

$$\nabla^2 V = 2B^T QB. \tag{3.2.4}$$

Exercises

1. If $x = (x_1, x_2)^T$ and $F(x) = \frac{1}{2}x^T Ax + b^T x$ show that

$$\nabla F = Ax + b \quad \text{and} \quad \nabla^2 F = A.$$

2. By generalising the result in question 1 for functions of more than two variables, show that (3.2.3) and (3.2.4) are the gradient and Hessian of the function in (3.2.1).

3. Using $\bar{r}$ and Q for the data in Table 1.2, calculate the gradient and Hessian of V in (3.2.1). Show that the Hessian V_{xx} has two positive eigenvalues.

4. Use (3.1.5), (3.1.6) to show that **Risk-Ret1** (section 1.3 in chapter 1) can be re-formulated as

$$\text{Minimize} \quad F = -\bar{r}^T(\alpha + Bx) + \rho(\alpha + Bx)^T Q(\alpha + Bx). \tag{3.2.5}$$

Show also that the gradient and Hessian of F in (3.2.5) are

$$\nabla F = -B^T \bar{r} + 2\rho B^T Q\alpha + 2\rho B^T QBx \tag{3.2.6}$$

$$\nabla^2 F = 2\rho B^T QB. \tag{3.2.7}$$

5. Evaluate the gradient and Hessian (3.2.6) and (3.2.7) when $\bar{r}$ and Q come from the data in Table 1.1 and $\rho = 100$.

6. As an alternative to the function (1.3.7) we could consider

$$F = \frac{V}{R} = \frac{y^T Qy}{\bar{r}^T y}. \tag{3.2.8}$$

Use (3.1.5), (3.1.6) to express (3.2.8) as a function of x, where

$$x_i = y_i, \; i = 1, \ldots, n-1 \quad \text{and} \quad y_n = 1 - \sum_{i=1}^{n-1} x_i.$$

Hence obtain expressions for the gradient and Hessian of F with respect to the independent variables $x_1, \ldots, x_{n-1}$.

3. Minimum risk for specified return

We showed in section 1.3 of chapter 1 how **Minrisk1** could be tackled via the related problem **Minrisk1m**. For the n-asset case, **Minrisk1m** can in turn be expressed, using (3.1.6), as the unconstrained minimization problem

$$\text{Minimize} \quad F = (\alpha + Bx)^T Q(\alpha + Bx) + \frac{\rho}{R_p^2}(\bar{r}^T\alpha - R_p + \bar{r}^T Bx)^2. \qquad (3.3.1)$$

In order to obtain expressions for the gradient and Hessian of the function F in (3.3.1) it will be convenient to write it as

$$F = V + \bar{\rho}(R - R_p)^2$$

where

$$V = (\alpha + Bx)^T Q(\alpha + Bx), \quad \bar{\rho} = \frac{\rho}{R_p^2} \text{ and } R = \bar{r}^T(\alpha + Bx).$$

Using subscript notation for gradients, the chain rule of differentiation implies

$$F_x = V_x + 2\bar{\rho}(R - R_p)R_x$$

where

$$V_x = 2B^T Q(\alpha + Bx) \quad \text{and} \quad R_x = B^T\bar{r}.$$

Similarly, using double subscripts to denote Hessian matrices,

$$F_{xx} = V_{xx} + 2\bar{\rho}(R - R_p)R_{xx} + 2\bar{\rho}R_x R_x^T.$$

We can see immediately that R_{xx} is the zero matrix since R_x is independent of x. Therefore we have

$$F_{xx} = V_{xx} + 2\bar{\rho}R_x R_x^T,$$

where

$$V_{xx} = 2B^T QB$$

and $R_x R_x^T$ defines a *rank-one* matrix whose (i, j)-th element is given by

$$(R_x R_x^T)_{i,j} = \frac{\partial R}{\partial x_i}\frac{\partial R}{\partial x_j}.$$

Exercises

1. If v is a 2-vector, show that the matrix $W = vv^T$ is symmetric. Show also that W has one zero eigenvalue. Generalise this result to show that, when v is an n-vector, $W = vv^T$ has at most one non-zero eigenvalue. (This is why W is called a rank-one matrix).

2. Calculate $\bar{r}$ and Q for the data in Table 3.1. If the target return R_p is 0.4% and $\rho = 100$, find the coefficients of the three-variable quadratic function (3.3.1) and hence determine its gradient and Hessian.

	Return % for					
	Period 1	Period 2	Period 3	Period 4	Period 5	Period 6
Asset 1	0.4	0.5	0.55	0.61	0.65	0.7
Asset 2	0.5	0.51	0.52	0.54	0.53	0.5
Asset 3	0.7	0.65	0.61	0.55	0.5	0.4
Asset 4	0.5	0.48	0.45	0.45	0.46	0.44

Table 3.1. Rates of return on four assets

Solving Minrisk1 by eliminating two variables

As in section 4 of chapter 1 we can solve **Minrisk1** by using (1.2.2) and (1.2.3) to express y_{n-1} and y_n in terms of $y_1, ..., y_{n-2}$. Thus, if we write

$$\gamma_1 = R_p - \sum_{i=1}^{n-2} \bar{r}_i y_i \quad \text{and} \quad \gamma_2 = 1 - \sum_{i=1}^{n-2} y_i \tag{3.3.2}$$

then (1.2.3) and (1.2.2) imply

$$\bar{r}_{n-1} y_{n-1} + \bar{r}_n y_n = \gamma_1 \tag{3.3.3}$$

$$y_{n-1} + y_n = \gamma_2. \tag{3.3.4}$$

Multiplying (3.3.4) by $\bar{r}_n$ and subtracting it from (3.3.3) gives

$$(\bar{r}_{n-1} - \bar{r}_n) y_{n-1} = \gamma_1 - \bar{r}_n \gamma_2 \tag{3.3.5}$$

and hence

$$y_{n-1} = \frac{\gamma_1 - \bar{r}_n \gamma_2}{\bar{r}_{n-1} - \bar{r}_n} = \tilde{\alpha}_{n-1} + \sum_{i=1}^{n-2} \tilde{\beta}_{n-1,i} y_i \tag{3.3.6}$$

$$\text{where} \quad \tilde{\alpha}_{n-1} = \frac{R_p - \bar{r}_n}{\bar{r}_{n-1} - \bar{r}_n} \quad \text{and} \quad \tilde{\beta}_{n-1,i} = \frac{\bar{r}_n - \bar{r}_i}{\bar{r}_{n-1} - \bar{r}_n}. \tag{3.3.7}$$

Moreover

$$y_n = \gamma_2 - y_{n-1} = \tilde{\alpha}_n + \sum_{i=1}^{n-2} \tilde{\beta}_{n,i} y_i \tag{3.3.8}$$

$$\text{where} \quad \tilde{\alpha}_n = 1 - \tilde{\alpha}_{n-1} \quad \text{and} \quad \tilde{\beta}_{n,i} = -(1 + \tilde{\beta}_{n-1,i}). \tag{3.3.9}$$

If we use these expressions to eliminate y_{n-1} and y_n the risk function can be considered as depending on $y_1, ..., y_{n-2}$ only. We define the n-vector $\tilde{\alpha}$ and the

$n \times (n-2)$ matrix $\tilde{B}$ by

$$\tilde{\alpha} = \begin{pmatrix} 0 \\ 0 \\ .. \\ 0 \\ \tilde{\alpha}_{n-1} \\ \tilde{\alpha}_n \end{pmatrix} \qquad \tilde{B} = \begin{pmatrix} 1 & 0 & \dots & 0 \\ 0 & 1 & \dots & 0 \\ .. & .. & .. & .. \\ 0 & 0 & \dots & 1 \\ \tilde{\beta}_{n-1,1} & \tilde{\beta}_{n-1,2} & \dots & \tilde{\beta}_{n-1,n-2} \\ \tilde{\beta}_{n,1} & \tilde{\beta}_{n,2} & \dots & \tilde{\beta}_{n,n-2} \end{pmatrix}.$$

Then, if x is the $(n-2)$-vector with $x_i = y_i,\ i = 1,\dots,n-2$ it follows that $y = \tilde{\alpha} + \tilde{B}x$ and the risk function (3.1.2) can be written as

$$V = y^T Q y = (\tilde{\alpha} + \tilde{B}x)^T Q (\tilde{\alpha} + \tilde{B}x).$$

Hence we can now pose **Minrisk1** as

$$\text{Minimize} \quad V = \tilde{\alpha}^T Q \tilde{\alpha} + 2\tilde{\alpha}^T Q \tilde{B} x + x^T \tilde{B}^T Q \tilde{B} x. \tag{3.3.10}$$

If x^* solves (3.3.10) then the optimal portfolio is given by $y_i^* = \tilde{\alpha} + \tilde{B}x^*$.

The reader can verify, by fairly simple algebra, that the gradient and Hessian of the function in (3.3.10) are given by

$$\nabla V = 2\tilde{B}^T Q \tilde{\alpha} + 2\tilde{B}^T Q \tilde{B} x. \tag{3.3.11}$$

$$\nabla^2 V = 2\tilde{B}^T Q \tilde{B}. \tag{3.3.12}$$

Exercise
Using $\bar{r}$ and Q for the data in Table 3.1, derive the two-variable risk function (3.3.10) when $R_p = 0.4\%$. Hence determine its gradient and Hessian.

4. The maximum return problem

The approach to the maximum return problem introduced in section 7 of chapter 1 can also be generalised to the case of n assets. The problem **Maxret1** is first reformulated as **Maxret1m** and then we can use (3.1.5) and the transformation $y = \alpha + Bx$ to obtain

$$\text{Minimize} \quad F = -\bar{r}^T(\alpha + Bx) + \frac{\rho}{V_a^2}[(\alpha + Bx)^T Q (\alpha + Bx) - V_a]^2. \tag{3.4.1}$$

To get expressions for the gradient and Hessian of F it is convenient to write

$$F = -R + \bar{\rho}(V - V_a)^2$$

$$\text{where} \quad R = \bar{r}^T(\alpha + Bx), \;\; \bar{\rho} = \frac{\rho}{V_a^2} \quad \text{and} \quad V = (\alpha + Bx)^T Q(\alpha + Bx).$$

The gradient of R is

$$R_x = B^T \bar{r}; \tag{3.4.2}$$

and the Hessian, R_{xx} is zero because R is linear in x. The gradient and Hessian of V are given by

$$V_x = 2B^T Q\alpha + 2B^T QBx$$

and

$$V_{xx} = 2B^T QB.$$

Using the chain rule, the gradient of the function in (3.4.1) can be written as

$$F_x = -B^T \bar{r} + 2\bar{\rho}(V - V_a)V_x. \tag{3.4.3}$$

The expression for its Hessian is

$$F_{xx} = 2\bar{\rho}(V_x V_x^T + (V - V_a)V_{xx}) \tag{3.4.4}$$

where the product $V_x V_x^T$ is a *rank-one* matrix whose i, j-th element is given by

$$(V_x V_x^T)_{ij} = \frac{\partial V}{\partial x_i}\frac{\partial V}{\partial x_j}.$$

Exercises

1. Using data from Table 1.1, derive the function (3.4.1) when $V_a = 0.003$ and $\rho = 1$; and hence determine its gradient and Hessian.

2. Explain why an alternative formulation of the maximum return problem could involve minimizing the function

$$F = \frac{(V - V_a)^2}{R} = \frac{(y^T Qy - V_a)^2}{\bar{r}^T y}. \tag{3.4.5}$$

Use (3.1.5) and the transformation $y = \alpha + Bx$ write F as function of x and then obtain expressions for F_x and F_{xx}.

3. Can you explain why the utility functions (3.2.8) and (3.4.5) might be unsatisfactory in practice for determining good investment solutions?

Taylor's Theorem

If we knew it all
for just a single moment
we'd hold the future

Mean Value Theorem

If we know in part
then, like stopped clocks, our forecasts
could be right - just once.

Chapter 4

UNCONSTRAINED OPTIMIZATION IN N VARIABLES

1. Optimality conditions

Conditions which characterise a minimum of an n-variable continuously differentiable function, $F(x_1, ..., x_n)$, are expressed in terms of the vector of first partial derivatives

$$g = (\frac{\partial F}{\partial x_1}, ..., \frac{\partial F}{\partial x_n})^T, \tag{4.1.1}$$

and the $n \times n$ matrix G of second partial derivatives whose (i, j)-th element is

$$G_{ij} = \frac{\partial^2 F}{\partial x_i \partial x_j}. \tag{4.1.2}$$

Definition The vector g in (4.1.1) is called the *gradient* and may also be written as ∇F (or sometimes as F_x).

Definition The matrix G given by (4.1.2) is known as the *Hessian* and may also be denoted by $\nabla^2 F$ or F_{xx}.

The Hessian matrix is always symmetric when F is a twice continuously differentiable function because then $\partial^2 F / \partial x_i \partial x_j = \partial^2 F / \partial x_j \partial x_i$. This will be the case for most of the problems we consider.

Definition A *positive definite* symmetric matrix is one which has all positive eigenvalues. Equivalently, a matrix A is positive definite if and only if

$$x^T A x > 0, \quad \text{for any } x \neq 0. \tag{4.1.3}$$

Definition If $F(x)$ is an n-variable function whose gradient and Hessian satisfy

$$g(x^*) = 0 \quad \text{and} \quad G(x^*) \text{ is positive definite.} \tag{4.1.4}$$

then the point x^* is a *local minimum* of $F(x)$.

It is the second of the *optimality conditions* (4.1.4) that distinguishes a minimum from a maximum (or any other stationary point). From a geometrical point of view, positive definiteness of G implies that the function is *convex* near the minimum. (Convexity is briefly discussed in section 2 below.)

It is the second derivative condition in (4.1.4) that ensures $F(x^*) < F(x)$ for all x in some, possibly small, region around x^*. For some functions $F(x)$ there may be several points x^* which satisfy (4.1.4). These are all *local* minima; and the one which gives the least value of F will be called the *global* minimum.

For simple functions we can use (4.1.4) directly. Consider the problem

$$\text{Minimize } F(x_1,x_2) = (x_1-1)^2 + {x_2}^3 - x_1x_2. \tag{4.1.5}$$

Setting the first partial derivatives to zero gives

$$2x_1 - 2 - x_2 = 0 \quad \text{and} \quad 3x_2^2 - x_1 = 0.$$

These equations have two solutions

$$(x_1,x_2) = (\frac{3}{4}, -\frac{1}{2}) \quad \text{and} \quad (x_1,x_2) = (\frac{4}{3}, \frac{2}{3}).$$

To identify the minimum we consider the Hessian

$$G = \begin{pmatrix} 2 & -1 \\ -1 & 6x \end{pmatrix}.$$

We can show quite easily that G is positive definite when $x_2 = \frac{2}{3}$ but not when $x_2 = -\frac{1}{2}$. Hence the minimum is at $(\frac{4}{3}, \frac{2}{3})$.

The problem **Minrisk0** can be solved by an analytical approach like the one in the previous example. Since the gradient of the function in (3.2.1) is

$$\nabla V = 2B^TQ\alpha + 2B^TQBx$$

the optimal values of the invested fractions $x = (y_1,...,y_{n-1})^T$ are found by solving the linear system of equations

$$B^TQBx = -B^TQ\alpha.$$

The remaining invested fraction y_n is then obtained as $1 - \sum_{i=1}^{n-1} y_i$.

We can obtain a similar algebraic solution for **Minrisk1** by the approach described in section 3 of chapter 1. In general, however, we cannot write down such expressions for the solutions when the function to be minimized is not quadratic and the first-order condition $g = 0$ does not yield equations which can be solved analytically. In particular, the solution of **Maxret1** involves the minimization of a quartic function. For such non-quadratic problems we may have to use iterative methods, just as in the one variable case. Many of these iterative methods are *gradient* techniques which – like the secant method or Newton's method – require the calculation of first (and sometimes second) derivatives. A number of such techniques will be described in the chapters which follow. We shall give a brief outline below of some direct search methods which use only function values.

Exercises

1. Prove that, when conditions (4.1.4) hold, $F(x^* + sp) > F(x^*)$ for any vector p, provided the scalar s is sufficiently small.

2. Use an analytical approach to find the minimum of the function (3.3.1) when its coefficients are obtained using the data from Table 3.1 and a target expected return $R_p = 0.4\%$.

3. Use an analytical approach to find the minimum of the utility function (3.2.5) when $\bar{r}$ and Q are determined from Table 1.1 and $\rho = 10$.

2. Visualising problems in several variables

It is easy to illustrate one-variable optimization problems using graphs of the objective function. For two-dimensional problems we can use *contour plots* and these can also give some insight into problems in higher dimensions. A contour plot for a function of two variables, x_1, x_2, shows curves in the (x_1, x_2)-plane along which the function has a constant value. (This is analagous to what is done on maps which show lines of constant altitude.)

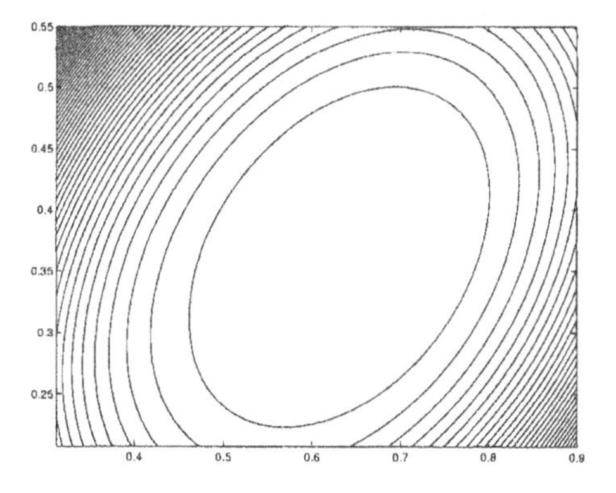

Figure 4.1. Convex contours of (3.4.1)

Figure 4.1 is a contour plot of the function (3.4.1) where $\bar{r}$ and Q are derived from the data in Table 1.1 and where $\rho = 0.1$ and $V_a = 0.001$. The minimum of the function is near the middle of the central oval region. The closed curves are contours whose function value increases towards the edges of the figure. The contours for (3.4.1) when $V_a = 0.001$ show that the function is *convex*, at least for the range of x_1, x_2 values illustrated. Convexity of a function means that *any* two points lying inside one of its contour lines can be joined by a straight line which also lies entirely inside the same contour line. Thus, if $F(a) < \hat{F}$ and $F(b) < \hat{F}$ then $F(a+\lambda(b-a)) < \hat{F}$ for $0 \leq \lambda \leq 1$.

Pictorially, a non-convex function has contour lines which "double-back on themselves". Examples appear in contour plots of (3.4.1) using data from Table 1.1 with $\rho = 0.1$ and $V_a = 0.002$ (Figure 4.2) and $V_a = 0.003$ (Figure 4.3).

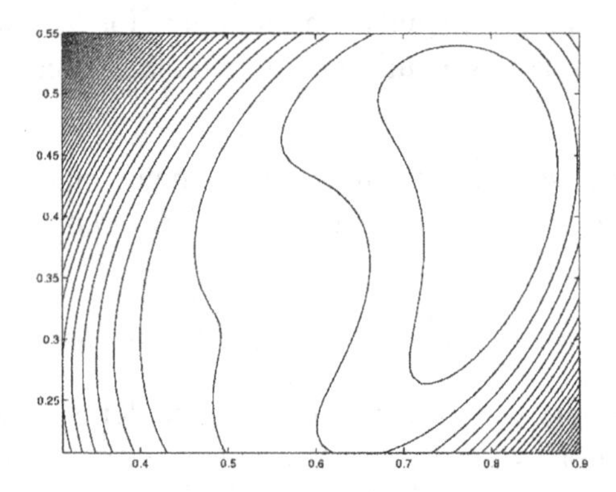

Figure 4.2. Nonconvex contours of (3.4.1)

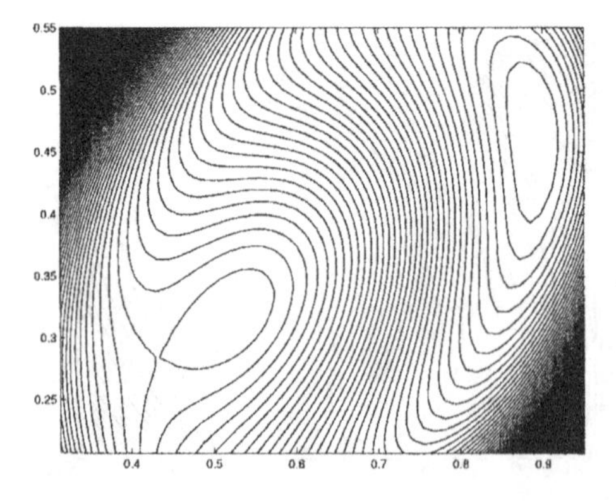

Figure 4.3. Multiple stationary points of (3.4.1)

If a function $F(x)$ is convex for all values of x then it must have a unique minimum. Non-convex functions, however, may have multiple stationary points. In Figure 4.3 there is a local maximum in the region of $x_1 = 0.5$, $x_2 = 0.33$ and a saddle point near $x_1 = 0.42$, $x_2 = 0.27$.

3. Direct search methods

We have already shown that it can be fairly straightforward to obtain gradients and Hessians of the expressions appearing in portfolio optimization problems. For this reason, the chapters which follow will focus chiefly on gradient methods of minimization. However there are also approaches which – like the bisection method in one variable – rely only on function values. These are usually referred to as *direct search* methods and are useful when we want to minimize functions which are not (or not easily) differentiable. For completeness, some of them will be briefly described below.

Obviously we could search for the least value of a function by evaluating it at all points on a "grid" of possible values of the variables; but clearly this is not very efficient. In some cases we might approximate the minimum more rapidly by sampling the function value at a sequence of "random" points, using statistical arguments to estimate the likelihood of finding the minimum in a certain number of trials. However, we shall confine ourselves to approaches that are more systematic.

Univariate search

Univariate search involves using a direct search method (like bisection) to do a sequence of one-dimensional minimizations of $F(x)$, first with respect to x_1, then with respect to x_2 and so on. In other words we search for the best point along each of the co-ordinate directions in turn. In general many such "cycles" of minimization for all n variables are needed; and though the method occasionally works quite well, it is not guaranteed to converge.

The Hooke and Jeeves method

This technique [4] uses the univariate search method just described, but follows it with a "pattern move". Thus, if $\tilde{x}$ and $\hat{x}$ are, respectively, the estimates of the minimum of $F(x)$ at the start and end of a univariate-search cycle then the pattern move involves a one-dimensional minimization of F along the direction $\hat{x} - \tilde{x}$. Thus we get an updated estimate of the form

$$\tilde{x} = \tilde{x} + \lambda(\hat{x} - \tilde{x})$$

for some scalar λ. The method continues in this way using univariate-search cycles followed by pattern moves.

The Nelder and Mead Simplex method
This method [5] is a more effective approach than univariate search. (It should not be confused with the Simplex method in Linear Programming.) We shall outline the approach for a two variable optimization problem.

We first make some initial estimates of the position of the minimum, and for a two-variable problem we take *three* initial points. These will define a starting *simplex*. (More generally, for a function of n variables, a simplex consists of $n+1$ points.) Suppose we label the *vertices* of the simplex as A, B, C and call the corresponding function values F_a, F_b, F_c. The vertex with the highest function value is said to be the *worst*; and this point must be replaced with a better one.

The basic move in the simplex method is *reflection*. A new trial point is obtained by reflecting the worst point in the centroid of the remaining vertices. This is a heuristic way of placing a new solution estimate in a region where function values are likely to be lower. Suppose, for instance, that $F_a > F_b > F_c$. Then the vertex A would be reflected in the centroid of vertices B and C. Let the new point be labelled as N with F_n as the associated function value. If $F_n < F_a$ the new point is an improvement on vertex A and a new simplex is defined by deleting the old worst point and renaming vertex N as A.

If the reflection gives a trial point N such that $F_n > F_a$ then the simplex method generates another trial point M so that $AM = \frac{2}{3}(AN)$. If $F_m < F_a$ then M becomes part of a new simplex; otherwise we must try a new vertex O where $AO = \frac{2}{3}(AM)$. This trial-and-error process is called *modified reflection* and its frequent use can indicate that the method is approaching a minimum. In these circumstances a *contraction* strategy may be used. This involves shrinking the current simplex towards its best point, so that, if B is the lowest vertex, a new simplex is obtained by halving the distances of all other vertices from B. The method stops when the simplex has shrunk below a certain size.

Using finite difference approximations to derivatives
One of the most common ways of minimizing $F(x)$ using function values only is by adapting gradient methods to use *finite-difference estimates* instead of accurate derivatives. Thus, for instance,

$$\frac{\partial F(x_1, x_2, .., x_n)}{\partial x_1} \approx \frac{F(x_1 + h, x_2, .., x_n) - F(x_1 - h, x_2, .., x_n)}{2h}$$

is the central-difference formula for first derivatives. Approximating derivatives in this way often works quite well, although a poor choice of the differencing step-size h can give inaccurate results and hinder the ultimate convergence to x^*. However the approach does rest on the implicit assumption that

F is actually differentiable and it may not be suitable for problems where F is non-smooth – i.e., has discontinuous derivatives.

DIRECT
In addition to the approaches outlined above we can also mention here another direct-search algorithm called DIRECT [6]. This will be described in a later chapter since it is a global, rather than simply a local optimization procedure.

Exercise
Do two cycles of univariate minimization on the function (4.1.5) starting from $x_1 = x_2 = 1$. What happens if a pattern move is added?

4. Optimization software & examples

We have already mentioned the `SAMPO` software in chapter 1. `SAMPO` will play an important part in the practical study of methods for n-variable optimization since it includes implementations of most of the techniques described in the chapters which follow. We shall use these implementations to compare the performance of different optimization techniques on realistic problems.

`SAMPO` can be downloaded from an `ftp` site (see appendix) and used to solve many of the exercises appearing in the rest of this book. Use of this fortran90 software is, however, not essential and many of the sample problems and exercises can be handled by any other optimization software which the reader can access – for instance, the `SOLVER` spreadsheet tool [7] in Microsoft Excel [8] or the optimization toolbox from `MATLAB` [9].

We will now describe some features of `SAMPO` in order to provide a background to the numerical results quoted in later chapters. In particular, we comment on the program `sample3` which solves problems **Minrisk1m** and **Maxret1m** by a number of different optimization methods. For all solution techniques, `sample3` uses the same standard starting guess for the invested fractions, namely

$$y_i = 1/n. \qquad (4.4.1)$$

It also employs a *standard* convergence test

$$||\nabla F(x)||_2 \le 10^{-5} \times \sqrt{n} \qquad (4.4.2)$$

for terminating the minimization iterations. (Note that both (4.4.1) and (4.4.2) are "self-adjusting" with n, the number of assets.) As well as the standard convergence rule (4.4.2), `sample3` allows us to choose *low* or *high* accuracy solutions defined, respectively, by

$$||\nabla F(x)||_2 \le 10^{-4} \times \sqrt{n} \quad \text{and} \quad ||\nabla F(x)||_2 \le 10^{-6} \times \sqrt{n}. \qquad (4.4.3)$$

One way of assessing the relative performance of two iterative minimization methods is to run them both on the same set of problems – using, of course, the same starting guesses for the variables and the same convergence test – and to compare the number of iterations they use. However, an iteration count is not the only measure of efficiency for optimization methods. The amount of computing effort needed to solve a problem also depends on the number of evaluations of the objective function (and perhaps its derivatives). Thus, when we quote results obtained with `SAMPO`, we shall typically state both the number of iterations performed and also the number of *calls* to the procedure which calculates $F(x)$ (together with the gradient and Hessian if these are used).

We can now say a few words about the examples used in this book. With `SAMPO` we can obtain data for a minimum-risk or maximum-return problem in one of three ways.
(a) A history of asset returns, as in Table 1.1, can be read in as a data file. This is useful for setting up small test problems, perhaps with special features.
(b) A history of real-life asset returns can be selected from a (small) database which is built in to `SAMPO`. This contains figures from trading on the London Stock Exchange during 2002 and 2003. With this option we can look at the actual performance of optimized portfolios. Data extracted from this built-in database will be denoted by `Real-m-n` to indicate an m-day history for the first n of the available assets.
(c) Synthetic – but fairly realistic – asset histories can be generated by making pseudo-random perturbations to the real-life data mentioned in *(b)*. With this feature we can easily generate some quite large problems – i.e. those involving several hundred assets – rather than the small-scale ones we have looked at so far. Large problems can pose particular challenges to minimization methods and we shall discuss these in a later chapter.
Using these ideas, we introduce our first set of test problems.

Problem T1 is based on the asset history in Table 4.1 and involves finding the minimum-risk portfolio to give a target return $R_p = 1.15\%$.

	Return % for period									
	1	2	3	4	5	6	7	8	9	10
Asset 1	1.2	1.3	1.4	1.5	1.1	1.2	1.1	1.0	1.0	1.1
Asset 2	1.3	1.0	0.8	0.9	1.4	1.3	1.2	1.1	1.2	1.1
Asset 3	0.9	1.1	1.0	1.1	1.1	1.3	1.2	1.1	1.0	1.1
Asset 4	1.1	1.1	1.2	1.3	1.2	1.2	1.1	1.0	1.1	1.2
Asset 5	0.8	0.75	0.65	0.75	0.8	0.9	1.0	1.1	1.1	1.2

Table 4.1. Rates of return on five assets over ten periods

The expected returns for the assets in Table 4.1 are given by

$$\bar{r} \approx (1.19,\ 1.13,\ 1.09,\ 1.15,\ 0.905)^T \tag{4.4.4}$$

and the variance-covariance matrix is

$$Q \approx \begin{pmatrix} 0.0249 & -0.0187 & -0.0011 & 0.0085 & -0.0219 \\ -0.0187 & 0.0321 & 0.0033 & -0.0035 & 0.0099 \\ -0.0011 & 0.0033 & 0.0109 & 0.0015 & 0.0035 \\ 0.0085 & -0.0035 & 0.0015 & 0.0065 & -0.0058 \\ -0.0219 & 0.0099 & 0.0035 & -0.0058 & 0.0307 \end{pmatrix}. \tag{4.4.5}$$

We can solve problem T1 via the formulation **Minrisk1m**. The result obtained will obviously depend on the value of ρ; but if ρ is sufficiently large to ensure that the actual return $R \approx R_p$ then the solution of problem T1 has invested fractions given approximately by

$$y_1 \approx 0.42,\ y_2 \approx 0.34,\ y_3 \approx 0.01,\ y_4 \approx 0.2,\ y_5 \approx 0.04 \tag{4.4.6}$$

with minimum risk $V \approx 3.44 \times 10^{-3}$.

Problem T2 involves the real-life asset data `Real-20-5` from the `SAMPO` database mentioned above. It seeks the minimum-risk portfolio which will yield a target return $R_p = 0.25\%$. The expected returns for the five assets are

$$\bar{r} \approx (-0.028,\ 0.366,\ 0.231,\ -0.24,\ 0.535)^T \tag{4.4.7}$$

and so the choice of R_p is reasonable since it lies comfortably within the range of elements of $\bar{r}$. The variance-covariance matrix for `Real-20-5` is

$$Q \approx \begin{pmatrix} 1.0256 & -0.4340 & 0.0202 & -0.1968 & -0.0311 \\ -0.4340 & 1.1049 & -0.0783 & 0.2347 & -0.1776 \\ 0.0202 & -0.0783 & 0.4328 & -0.1236 & -0.1895 \\ -0.1968 & 0.2347 & -0.1236 & 8.0762 & 1.0093 \\ -0.0311 & -0.1776 & -0.1895 & 1.0093 & 2.9007 \end{pmatrix}. \tag{4.4.8}$$

(We note that real-life asset histories can yield negative values of expected returns and also lead to much larger values in the elements of Q than those given by artificial datasets like Table 4.1.)

The invested fractions which solve problem T2 are given approximately by

$$y_1 \approx 0.2,\ y_2 \approx 0.27,\ y_3 \approx 0.41,\ y_4 \approx -0.003,\ y_5 \approx 0.11 \tag{4.4.9}$$

with minimum risk $V \approx 0.1426$.

Problem T3 uses the data in Table 1.1 and involves finding the maximum return that can be obtained for an acceptable risk $V_a = 0.003$. The expected returns for the three assets in Table 1.1 are

$$\bar{r} \approx (1.283,\ 1.117,\ 1.083)^T \tag{4.4.10}$$

and the variance-covariance matrix is

$$Q \approx \begin{pmatrix} 0.0181 & -0.0281 & -0.00194 \\ -0.0281 & 0.0514 & 0.00528 \\ -0.00194 & 0.00528 & 0.0147 \end{pmatrix}. \tag{4.4.11}$$

The choice of V_a in a maximum-return problem can be related to the solution of the basic minimum-risk problem **Minrisk0**. For the assets in Table 1.1, the least possible risk V_{min} is about 0.00111 and so problem T3 seeks the best return that can be obtained if we are willing to accept a risk of about $3V_{min}$.

The solution to problem T3 can be obtained by solving **Maxret1m** and is given, approximately, by

$$y_1 \approx 0.86,\ y_2 \approx 0.45,\ y_3 \approx -0.31 \tag{4.4.12}$$

yielding an expected return $R \approx 1.27\%$.

Problem T4 is a maximum-return example using the dataset `Real-20-5` which gives the $\bar{r}$ and Q values (4.4.7), (4.4.8). It involves finding the maximum return that can be obtained for an acceptable risk $V_a = 0.15$ (which, for these five assets, is about $1.08V_{min}$). The solution is given by

$$y_1 \approx 0.18,\ y_2 \approx 0.28,\ y_3 \approx 0.42,\ y_4 \approx -0.01,\ y_5 \approx 0.13 \tag{4.4.13}$$

with an expected return $R \approx 0.267\%$.

Chapter 5

THE STEEPEST DESCENT METHOD

1. Introduction

The *steepest descent* method is the simplest of the gradient methods for optimization in n variables. It can be justified by the following geometrical argument. If we want to minimize a function $F(x)$ and if our current trial point is x_k then we can expect to find better points by moving away from x_k along the *search direction* which causes F to decrease most rapidly. This is the direction of the negative gradient. To use a geographical illustration: suppose we are walking on a hillside and wish to get to the bottom of the valley. If we keep moving straight down the line of greatest slope we shall, in general, cover less ground before reaching the valley floor than if we follow an oblique path across the valley wall.

A formal description of the steepest descent method appears below. Here, and in what follows, subscripts on vectors will be used to denote iteration numbers. On occasions when we need to refer to the i-th element of a vector x_k we shall use double-subscript notation x_{k_i}.

Steepest Descent with perfect line search

Choose an initial estimate, x_0, for the minimum of $F(x)$.
Repeat for $k = 0, 1, 2, ...$
set $p_k = -\nabla F(x_k)$
calculate s^* to minimize $\varphi(s) = F(x_k + sp_k)$
set $x_{k+1} = x_k + s^* p_k$
until $||\nabla F(x_{k+1})||$ is sufficiently small.

The one-dimensional minimization in this algorithm can be performed using methods discussed in chapter 2.

It should be said at once that the steepest descent algorithm is not a particularly efficient minimization method. (The simple strategy of proceeding along the negative gradient works well for functions with near-circular contours; but practical optimization problems often involve functions with narrow, curving valleys and these need a more sophisticated approach.) However we shall consider it at some length because it introduces a pattern which is common to many optimization methods: namely that an iteration consists of two parts – the choice of a *search direction* (p_k) followed by a *line search* to determine a suitable step-size s^*.

2. Line searches

Definition A line search which chooses s^* to minimize $\varphi(s) = F(x_k + sp_k)$ is said to be *perfect* or *exact*.

Definition A *weak* or *inexact* line search is one which accepts any value of s such that $F(x_k + sp_k) - F(x_k)$ is negative and bounded away from zero.

A perfect line search gives the greatest possible reduction in F along the search direction. However, as we shall see later, it may be computationally expensive to do an accurate minimization of $\varphi(s)$ on every iteration. Hence weak searches are often preferred in practice. A convergence proof for the steepest descent algorithm with a weak line search will be given later in this chapter.

Line searches play an important part in optimization and we need to study them more closely. If p denotes any search direction and if we write

$$\varphi(s) = F(x_k + sp) \tag{5.2.1}$$

then, using a Taylor expansion,

$$\varphi(s) = F(x_k) + sp^T \nabla F(x_k) + \frac{s^2}{2} p^T \nabla^2 F(x_k) p + \dots$$

and so

$$\frac{d\varphi}{ds} = p^T \nabla F(x_k) + sp^T \nabla^2 F(x_k) p + \dots$$

But

$$\nabla F(x_k + sp) = \nabla F(x_k) + s\nabla^2 F(x_k) p + \dots$$

and so

$$\frac{d\varphi}{ds} = p^T \nabla F(x_k + sp). \tag{5.2.2}$$

(We can also derive this relationship by using the chain rule.) From (5.2.2) we deduce that the initial slope, as we move away from x_k along the search direction p, is given by $p^T \nabla F(x_k)$.

Definition The vector p is a *descent direction* with respect to the function $F(x)$ at the point x_k if it satisfies the condition

$$p^T \nabla F(x_k) < 0. \qquad (5.2.3)$$

If (5.2.3) holds then p is a suitable search direction for an iteration of a minimization algorithm which begins at x_k.

Proposition If s^* is the step which minimizes $\varphi(s)$ defined by (5.2.1) then

$$p^T \nabla F(x_k + s^* p) = 0. \qquad (5.2.4)$$

Proof The result follows on putting $d\varphi/ds = 0$ on the left of (5.2.2).

Condition (5.2.4) means that a perfect line search terminates at a point where the gradient vector is orthogonal to the direction of search.

A steepest descent example

We now apply steepest descent to the function

$$F(x) = (x_1 - 1)^2 + x_2^3 - x_1 x_2,$$

whose gradient is

$$g = (2x_1 - 2 - x_2,\ 3x_2{}^2 - x_1)^T.$$

If the starting point is $x_0 = (1,\ 1)^T$ then $F_0 = 0$ and $g_0 = (-1,\ 2)^T$. The first search direction is $p_0 = -g_0$ and the first iteration gives

$$x = x_0 + sp_0 = (1,\ 1)^T + s(1,\ -2)^T = (1+s,\ 1-2s)^T.$$

Hence

$$\varphi(s) = F(x_0 + sp_0) = s^2 + (1-2s)^3 - (1+s)(1-2s). \qquad (5.2.5)$$

We can use (5.2.3) to confirm that p_0 is a descent direction because

$$p_0^T g_0 = (1,\ -2)\begin{pmatrix} -1 \\ 2 \end{pmatrix} = -5.$$

In order to find the steplength s^* which minimizes $\varphi(s)$ we solve $d\varphi/ds = 0$. For the present problem this gives a quadratic equation which we can deal with easily. In general, however, s^* would have to be calculated by an iterative procedure such as the bisection method. From (5.2.5) we obtain

$$\frac{d\varphi}{ds} = 2s - 6(1-2s)^2 - (1-2s) + 2(1+s) = -24s^2 + 30s - 5.$$

On solving $24s^2 - 30s + 5 = 0$ we find the smaller root is $s^* \approx 0.1980$. Hence a perfect search will give the new point $x = (1.198,\ 0.604)^T$. (The larger root of the quadratic equation for s corresponds to a *maximum* of φ.)

A second steepest descent iteration from $x_1 = (1.198,\ 0.604)^T$ will use

$$p_1 = -g_1 = (0.208,\ 0.1036)^T$$

as the search direction. The new point will be of the form

$$x_2 = (1.198 + 0.208s,\ 0.604 + 0.1036s)^T$$

and s can again be chosen by a perfect line search. Continuing in this way we can expect that a minimum will be found if enough iterations are performed.

Exercises

1. In the first iteration of the worked example in this section, show that the same value of s^* would be obtained by solving the equation $p_0^T g = 0$ where the gradient g is calculated at $x = (1+s,\ 1-2s)^T$.

2. Perform another iteration of the steepest descent method with perfect line searches applied to (4.1.5) following on from the point $x_1 = (1.198,\ 0.604)^T$.

3. Show that the steepest descent method with perfect line searches generates successive search directions that are orthogonal.

3. Convergence of the steepest descent method

Experience shows that methods which use perfect line searches may not make significantly better overall progress than those which take steps which do not precisely minimize the line search function $\varphi(s)$. The following result demonstrates that the steepest descent method with a weak line search will converge to a stationary point of a function.

Proposition Let $F(x)$ be a function which is twice continuously differentiable and bounded below; and let its Hessian matrix be bounded, so that for some positive scalar M,

$$z^T \nabla^2 F(x) z \leq M||z||^2$$

for any vector z. Then a sequence of steepest descent iterations

$$x_{k+1} = x_k - \frac{1}{M}\nabla F(x_k)$$

(i.e., which use a constant step size $s = M^{-1}$) will produce a sequence of points x_k such that $||\nabla F(x_k)|| \rightarrow 0$ as $k \rightarrow \infty$.

Proof Suppose the statement is false and that, for some positive ε,

$$||\nabla F(x_k)|| > \varepsilon \quad \text{for all} \quad k.$$

Now consider a typical iteration starting from a point x where $p = -\nabla F(x)$ and $x^+ = x + sp$. By the Mean Value Theorem, for some ξ between x and x^+,

$$F^+ - F = sp^T \nabla F(x) + \frac{1}{2} s^2 p^T \nabla^2 F(\xi) p. \qquad (5.3.1)$$

Hence, writing g for $\nabla F(x)$,

$$F^+ - F = -\frac{g^T g}{M} + \frac{g^T \nabla^2 F(\xi) g}{2M^2} \leq -\frac{g^T g}{M} + \frac{g^T g}{2M},$$

using the bound on $\nabla^2 F$. Now by the assumption at the start of the proof we have, on every iteration

$$F^+ - F \leq -\frac{\varepsilon^2}{2M}. \qquad (5.3.2)$$

But if this holds for an infinite number of steps it contradicts the fact that $F(x)$ is bounded below; and hence our initial assumption must be false and there exists an integer K such that $||\nabla F(x_k)|| \leq \varepsilon$ for all $k > K$.

The above result does not relate to an algorithm which is either practical or efficient. We would not in general be able to determine the constant M; and, even if we could, the step-size $s = M^{-1}$ would usually be much less than the perfect step and convergence would be slow. However we can use the same *reductio ad absurdum* approach to show the convergence of the steepest descent algorithm as stated in section 1 of this chapter.

Corollary If the function $F(x)$ satisfies the conditions of the preceding proposition then the steepest descent algorithm with perfect line searches produces a sequence of points x_k such that $||\nabla F(x_k)|| \rightarrow 0$ as $k \rightarrow \infty$.

Proof This result follows because, with a perfect line search, the decrease in function value obtained on every iteration must be at least as good as that given by the bound (5.3.2). Therefore it would still imply a contradiction of F being bounded below if the iterations did not approach a stationary point.

The rate of convergence of steepest descent

The fact that an algorithm can be proved to converge does not necessarily imply that it is a good method. Steepest descent, whether using perfect or weak line searches, is not usually to be recommended in comparison with the algorithms introduced in later chapters. This is because its rate of convergence can be slow, as shown in the next example. Consider the problem

$$\text{Minimize } F(x) = \frac{1}{2}(x_1^2 + Cx_2^2).$$

Then $\nabla F = (x_1, Cx_2)^T$ and F has a minimum at $x^* = (0, 0)^T$. (A simple sketch shows that the contours are ellipses.)

If we choose $x_0 = (1, C^{-1})^T$ as a starting point then $p_0 = (-1, -1)^T$. Thus the next iterate will be of the form

$$x_1 = (1 - s, C^{-1} - s)^T. \quad (5.3.3)$$

A perfect line search finds s so that $p_0^T g_1 = 0$, i.e.

$$(-1, -1)\begin{pmatrix} 1-s \\ 1-Cs \end{pmatrix} = -1 + s - 1 + Cs = 0$$

and so $s = 2(1+C)^{-1}$. Substituting in (5.3.3) we get the new point

$$x_1 = K(1, -C^{-1})^T, \text{ where } K = \frac{(C-1)}{(C+1)}. \quad (5.3.4)$$

It follows from (5.3.4) that $||x_1|| = K||x_0||$ and so the error after the first iteration is K times the error at the starting point. In a similar way we can show

$$||x_2|| = K||x_1|| = K^2||x_0||. \quad (5.3.5)$$

In the special case when $C = 1$ (when F has circular contours) steepest descent performs well, since K is zero and the solution is found in one iteration. However, (5.3.5) shows that, for larger values of C, the solution error decreases by a constant factor K. Moreover, K is close to 1 for quite moderate values of C. For instance, $K \approx 0.82$ when $C = 10$ and so about 60 iterations would be needed to reduce $||x||$ to around 10^{-5}. Convergence would be yet slower for $C = 100$. This example demonstrates the potential inefficiency of the steepest descent approach. It also illustrates a general property of the steepest descent method, which we state without proof.

Proposition If $F(x)$ is a function for which the steepest descent algorithm converges to a stationary point x^*, then there exists an integer $\bar{k}$ and a positive real constant $K(< 1)$ such that, for $k > \bar{k}$,

$$||x_{k+1} - x^*|| < K||x_k - x^*||.$$

This means that the steepest descent method generally displays *linear* convergence near the solution, with the errors in the approximate minima, x_k, often decreasing by a constant factor on every iteration.

4. Numerical results with steepest descent

Performance of the steepest descent method can be illustrated by results from the program `sample3` discussed at the end of chapter 4. In this section we only consider the steepest descent method with perfect line searches, denoted by SDp. Table 5.1 gives information about solutions to problems T1 – T4 obtained by applying SDp to **Minrisk1m** or **Maxret1m** with different values of ρ and using the standard convergence test (4.4.2). Specifically we show the numbers of iterations and function evaluations required, together with the values of expected return and minimum risk for the calculated portfolio. Note that, as ρ increases, these return and risk values get closer to the target and solution values given in the problem descriptions in section 4 of chapter 4.

Problem	itns/fns	R	V
T1 $\rho = 1$	245/492	1.12%	1.8×10^{-3}
T1 $\rho = 10$	279/560	1.14%	3.1×10^{-3}
T1 $\rho = 100$	153/383	1.15%	3.4×10^{-3}
T2 $\rho = 1$	222/445	0.24%	0.1406
T2 $\rho = 10$	191/479	0.25%	0.1423
T2 $\rho = 100$	483/1209	0.25%	0.1425
T3 $\rho = 1$	143/610	1.27%	3.1×10^{-3}
T3 $\rho = 10$	1799/6529	1.27%	3.0×10^{-3}
T3 $\rho = 100$	fails	-	-
T4 $\rho = 1$	112/400	0.287%	0.1638
T4 $\rho = 10$	898/3198	0.270%	0.1519
T4 $\rho = 100$	811/3408	0.267%	0.1502

Table 5.1. SDp solutions for problems T1 – T4

Table 5.1 shows that SDp may need to perform a large number of iterations before the convergence test (4.4.2) is satisfied. On problem T3 (with $\rho = 100$) convergence has not occurred after 5000 iterations and so this case is recorded as a failure. For most of the problems in Table 5.1, the numbers of iterations and/or function calls increase with ρ.

Table 5.2 shows the performance of SDp with the low-accuracy convergence test (4.4.3) as well as the standard one. We see that SDp can take between 100 and 300 iterations just to reduce the gradient norm from $O(10^{-4})$ to $O(10^{-5})$. Hence the slow ultimate convergence of steepest descent, discussed in the preceding section, is not merely a theoretical difficulty but does occur in practice.

Problem	itns/fns (low)	itns/fns (standard)
T1 $\rho = 10$	81/164	279/560
T2 $\rho = 100$	355/889	483/1209
T3 $\rho = 10$	1485/5587	1799/6529
T4 $\rho = 100$	675/3000	811/3408

Table 5.2. Ultimate convergence of SDp on problems T1 – T4

Exercises (To be solved using `sample3` or other suitable software.)
1. Use the steepest descent method to calculate the efficient frontier for the dataset `Real-20-5`, considering R_p in the range $0.1\% \leq R_p \leq 0.5\%$. (The $\bar{r}$ and Q values for this problem are given by (4.4.7) and (4.4.8)).

2. How would the solution of problem T1 change if we only used the last seven time periods in the asset history?

3. Does the performance of the steepest descent method on problems T1 – T4 change if we use $y_1 = y_2 = ... = y_{n-1} = 0,\ y_n = 1$ as the starting guess?

4. Make a comparison table showing how the convergence tolerance affects numbers of iterations and function calls needed by the steepest descent method on the problems solved in question 1.

5. Wolfe's convergence theorem

The steepest descent method introduces some important ideas which are common to many other minimization techniques. These are (*i*) the choice of a *search direction*, p, to satisfy the descent property (5.2.3); and (*ii*) the use of a *line search* to ensure that the step, s, along p decreases the function. Optimization techniques differ mainly in the way that p is calculated on each iteration and, to a lesser extent, in how the step length, s, is obtained.

Wolfe's Theorem [10] gives precise conditions on p and s which guarantee convergence of any minimization algorithm. We now define these *Wolfe conditions* with x_k denoting an estimate of the minimum of $F(x)$ and $g_k = \nabla F(x_k)$.

Definition The *first Wolfe condition* is a stronger form of (5.2.3), namely

$$p^T g_k \leq -\eta_0 ||p||\ ||g_k||, \tag{5.5.1}$$

where η_0 is a small positive constant, typically $\eta_0 = 0.01$.
If (5.5.1) holds then θ, the angle between p and $-g_k$, is such that $\cos\theta$ is positive and bounded away from zero. In other words $-\frac{\pi}{2} < \theta < \frac{\pi}{2}$.

Before stating the other Wolfe conditions, we introduce notation to help us visualise the line-search function $\varphi(s) = F(x_k + sp)$. If s^* denotes the step to the minimum of φ then, in general, there is a step $\bar{s} > s^*$ such that $F(x_k + \bar{s}p) = F(x_k)$. On a quadratic function we can show that $\bar{s} = 2s^*$ (Exercise 2, below). The purpose of the next two Wolfe conditions is to define an acceptable step s as being one which is not too close to $\bar{s}$ or to zero.

Definition The *second Wolfe condition* is

$$F(x_k + sp) - F(x_k) \leq \eta_1 \, s \, p^T g_k, \tag{5.5.2}$$

for some constant η_1, such that $0.5 > \eta_1 > 0$ (typically $\eta_1 = 0.1$).
Condition (5.5.2) ensures that the step taken produces a non-trivial reduction in the objective function – i.e., that s is bounded away from $\bar{s}$.

Definition The *third Wolfe condition* is

$$|F(x_k + sp) - F(x_k) - s \, p^T g_k| \geq \eta_2 |s \, p^T g_k|, \tag{5.5.3}$$

where η_2 is a constant such that $0.5 > \eta_2 > 0$.
The inequality (5.5.3) ensures s is bounded away from zero by requiring the *actual* decrease in F to be bounded away from the linear *predicted* reduction.

These conditions are used in Wolfe's Theorem. The ideas in the proof (not given here) are similar to those used in the steepest descent convergence proof.

Wolfe's Theorem [10]
If $F(x)$ is bounded below and has bounded second derivatives then any minimization algorithm which satisfies (5.5.1) – (5.5.3) on a regular subsequence of iterations (and does not allow F to increase) will terminate in a finite number of iterations at a point where $||\nabla F(x)||$ is less than any chosen positive tolerance.

We can show the significance of the second and third Wolfe conditions if we introduce the ratio

$$D(s) = \frac{F(x_k + sp) - F(x_k)}{s \, p^T g_k}. \tag{5.5.4}$$

Clearly $D(\bar{s}) = 0$ and we can also show that $D(s) \to 1$ as $s \to 0$. If $F(x)$ is a quadratic function then $D(s)$ decreases linearly from 1 to 0 as s increases from 0 to $\bar{s}$. In particular $D(s^*) = 0.5$. This is illustrated in Figure 5.1.

The second Wolfe condition is equivalent to the requirement that $D(s) \geq \eta_1$. Similarly, the third Wolfe condition holds if $D(s) \leq 1 - \eta_2$. The vertical dashed lines in Figure 5.1 define the acceptable range for s when $\eta_1 = \eta_2 = 0.1$.

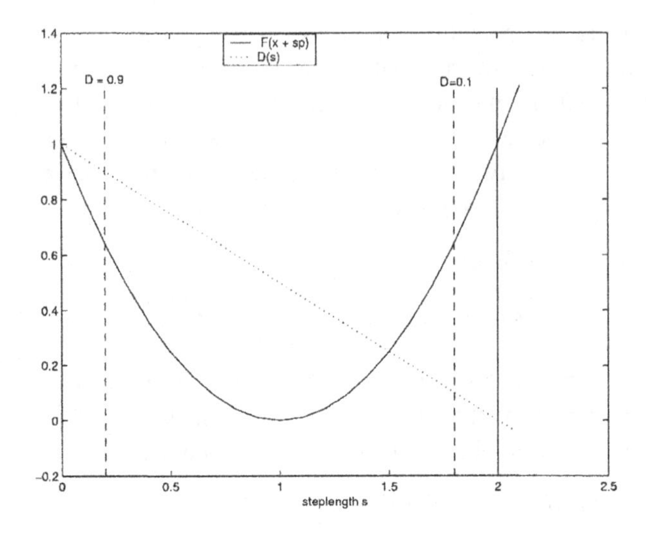

Figure 5.1. Wolfe conditions on a quadratic function

If F(x) is non-quadratic then $D(s)$ will not be linear. However, if F is convex, D will still lie in the range $1 \geq D(s) \geq 0$ when $0 \leq s \leq \bar{s}$. Figure 5.2 illustrates the acceptable range for s on a cubic function when $\eta_1 = \eta_2 = 0.1$. Note that the left- and right-hand excluded regions are not now the same size.

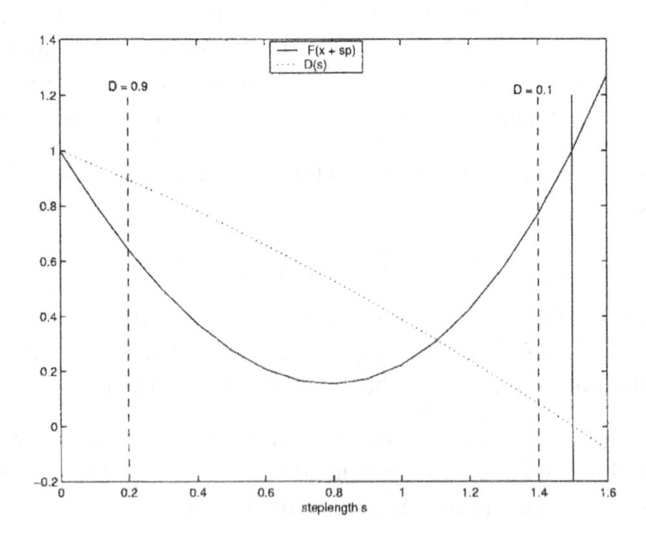

Figure 5.2. Wolfe conditions on a non-quadratic function

If $F(x)$ is non-convex then the ratio $D(s)$ may exceed 1 in the range $0 \leq s \leq \bar{s}$. This can be seen in Figure 5.3, where the function has slight negative curvature near to $s = 0$. The third Wolfe condition – that the step s must not be too close to zero – can then be expressed as

$$|1 - D(s)| \geq \eta_2.$$

Figure 5.3 shows the acceptable range for s when $\eta_1 = \eta_2 = 0.1$.

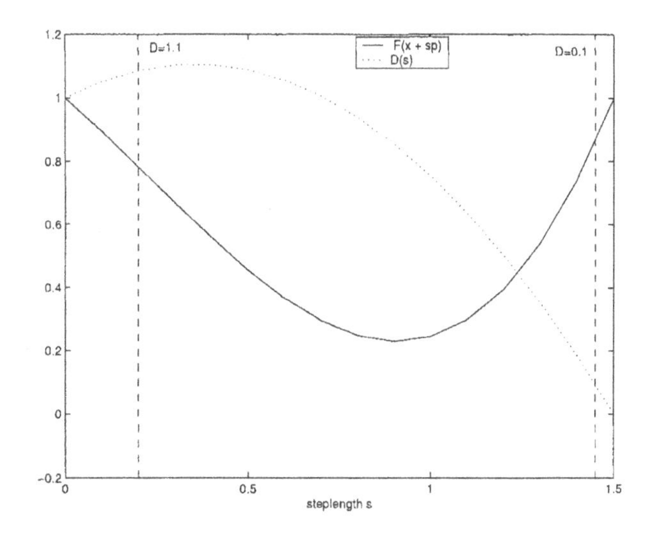

Figure 5.3. Wolfe conditions on a non-convex non-quadratic function

The ratio (5.5.4) is used in the Armijo line search procedure which is described in the next sub-section.

Exercises

1. Prove that search directions in the steepest descent method satisfy Wolfe Condition 1.

2. Using the notation from the paragraph preceding (5.5.2), prove that $\bar{s} = 2s^*$ when F is quadratic.

3. An alternative form of the third Wolfe condition is

$$|p^T g(x_k + sp)| \leq \eta_4 |p^T g(x_k)| \tag{5.5.5}$$

for some constant $\eta_4 (1 > \eta_4 \geq 0)$. Explain why this causes s to be bounded away from zero.

4. If $F(x)$ is quadratic, prove that (5.5.4), implies $D(s^*) = 0.5$.

The Armijo line search

Conditions (5.5.2) and (5.5.3) will certainly be be fulfilled if s minimises the line search function $\varphi(s)$ in (5.2.1). However, they also justify the use of a weak line search. This could be implemented by terminating an exact search before it converges. However, a simpler form of weak search, based on the second and third Wolfe conditions, is called the *Armijo Technique* [11] and can be described in terms of the ratio $D(s)$ defined by (5.5.4).

The Armijo Line Search Technique

Let p be a search direction satisfying (5.5.1)
Choose constants $C > 1$ and $c < 1$
Set $s = 1$ and $s_{min} = 0$
Repeat for $j = 0, 1, 2, \ldots$
If $|1 - D(s)| \geq \eta_2$ then exit; else set $s_{min} = s; \quad s = Cs$
Repeat for $k = 0, 1, 2, \ldots$
set $s = s_{min} + c(s - s_{min})$
until $D(s) \geq \eta_1$

This search first ensures that s is "big enough" and then modifies s if necessary to produce a decrease in F consistent with (5.5.2). The second, step-reducing, phase must not cause a violation of (5.5.3).

There is a modified form of Armijo search which can perform exact minimizations when F is quadratic by making use of the fact that $D(s^*) = 0.5$.

A Modified Armijo line search

Let p be a search direction satisfying (5.5.1)
Choose constants $C > 1$ and $c < 1$. Set $s = 1$ and $s_{min} = 0$.
Repeat for $j = 0, 1, 2, \ldots$
if $|1 - D(s)| \geq \eta 2$ then exit; else $s_{min} = s; \quad s = \min(Cs, \frac{0.5s}{1-D(s)})$
Repeat for $k = 0, 1, 2, \ldots$
set $s = \max(s_{min} + c(s - s_{min}), \frac{0.5s}{1-D(s)})$
until $D(s) \geq \eta_1$

This modified Armijo search uses linear interpolation or extrapolation to estimate a value of s such that $D(s) = 0.5$ (but using the constants C and c to prevent excessively large or small corrections). This means that the algorithm can do a perfect line search when φ is quadratic.

For non-quadratic functions, the Armijo search with $\eta_1, \eta_2 \approx 0.5$ will (usually) give a better estimate of the line minimum than one which uses $\eta_1, \eta_2 \approx 0.1$.

However, in order to perform a perfect line search on non-quadratic functions we cannot rely on an Armijo search by itself. We can use it to obtain an approximation to the line minimum, but we must then switch to a one-dimensional minimization technique (such as the secant method) in order to locate a point where the directional derivative $\varphi'(s) = p^T g$ is close enough to zero.

Exercise
Calculate the point reached on the first iteration of the example in section 2 of this chapter if the line search is done by the Armijo method with $\eta_1 = \eta_2 = 0.1$, using $s = 1$ as the first trial step. What would be the point reached if the modified Armijo search were used instead?

6. Further results with steepest descent

In this section we quote results from `sample3` to compare performance of the weak linesearch version of steepest descent, `SDw`, with that of `SDp` on problems T1 -T4. The particular values of ρ used in **Minrisk1m** (or in **Maxret1m**) has been chosen so that the target return R_p (or the acceptable risk V_a) is achieved to within $\pm 1\%$. The entries in Table 5.3 are numbers of iterations and function calls needed to satisfy the standard stopping rule (4.4.2).

	T1 $\rho = 100$	T2 $\rho = 50$	T3 $\rho = 2$	T4 $\rho = 50$
`SDp`	153/383	277/694	114/563	2985/10424
`SDw`	157/316	277/694	115/235	1047/3130

Table 5.3. Performance of `SDp` and `SDw` on problems T1 – T4

On problems T1 and T3, `SDw` takes more iterations than `SDp` but the weak line search uses fewer function calls per iteration and hence `SDw` seems more economical overall. On problem T4, `SDw` does substantially better than `SDp` in terms of both iteration count and function evaluations. The identical behaviour of the two versions on problem T2 occurs because the modified Armijo search can locate the one-dimensional minimum of a quadratic function if the initial trial step is close to being optimal. This appears to have been the case on every iteration during the solution of this example. Note that, on the non-quadratic problems T3 and T4, `SDp` typically uses many more function calls per linesearch than it does on problems T1 and T2.

Exercises (To be solved using `sample3` or other suitable software.)
1. Solve **Maxret1m** for the data in Table 1.1 with a target risk of $V_a = 0.0025$ using the method of steepest descent and observe how the performance varies as ρ changes.

2. Using the data in Table 4.1, apply steepest descent to minimize the function (3.4.1) with $V_a = 0.002$ and with various ρ betwen 0.1 and 1. Comment on the changes in the optimal portfolio.

3. For $0.14 \leq V_a \leq 0.2$, plot return against acceptable risk V_a at the solutions of **Maxret1m** obtained with dataset `Real-20-5` for which $\bar{r}$ and Q are given by (4.4.7) and (4.4.8). Comment on the result from minimizing **Maxret1m** when $V_a = 0.13$. (You should try several values of ρ when investigating this problem.)

4. Use the steepest descent method to minimize the function (3.2.5) for the data in Table 4.1. Hence calculate (and then plot) expected return against risk for ρ in the range $100 \leq \rho \leq 1000$.

5. By modifying `sample3` or otherwise write a procedure to minimize

$$F = \frac{V}{R} = \frac{y^T Q y}{\bar{r}^T y}$$

expressed as a function of $y_1,..,y_{n-1}$ only. Apply the procedure to problems T1 – T4. (**Hint**: for the necessary mathematics, see the exercises at the end of section 2.)

6. Show that it is not invariably the case that perfect line searches are less efficient than weak ones by using `SDp` and `SDw` to solve **Maxret1m** for the data in Table 3.1 with $R_p = 0.5\%$ and $\rho = 100$.

Chapter 6

THE NEWTON METHOD

1. Quadratic models and the Newton step

The steepest descent algorithm performs badly on the example in section 3 of chapter 5 chiefly because it uses no second derivative information. More effective methods are based on the properties of a *quadratic function*, $Q(x)$ which has a constant Hessian matrix. Let

$$Q(x) = \frac{1}{2}x^T Ax + b^T x + c. \tag{6.1.1}$$

Its gradient is

$$\nabla Q(x) = Ax + b \tag{6.1.2}$$

and we can find a stationary point of (6.1.1) by solving the linear equations

$$Ax = -b. \tag{6.1.3}$$

The solution will be a minimum if the Hessian matrix A is positive definite. On the other hand, it will be a maximum if A is negative definite or a saddle point if A is indefinite. If A is non-singular then (6.1.1) has a *unique* stationary point.

We can apply these ideas to the minimization of a general function, $F(x)$. Suppose x_k is an estimate of the minimum of $F(x)$ and that $g_k = \nabla F(x_k)$, $G_k = \nabla^2 F(x_k)$. We can write the truncated Taylor series

$$F(x_k + p) \approx Q(p) = F(x_k) + p^T g_k + \frac{1}{2}(p^T G_k p) \tag{6.1.4}$$

and

$$\nabla F(x_k + p) \approx \nabla Q(p) = g_k + G_k p. \tag{6.1.5}$$

Therefore, *if G_k is positive definite*, a prediction of the step from x_k to the minimum of F is

$$p = -G_k^{-1} g_k. \tag{6.1.6}$$

This leads to the following algorithm.

The Newton Method

Choose x_0 as an initial estimate of the minimum of $F(x)$
Repeat for $k = 0, 1, 2, \ldots$
Set $g_k = \nabla F(x_k)$, $G_k = \nabla^2 F(x_k)$.
if G_k is positive definite then obtain p_k by solving $G_k p_k = -g_k$
else set $p_k = -g_k$
Find s so $F(x_k + sp_k)$ satisfies (5.5.2), (5.5.3) for some η_1, η_2
Set $x_{k+1} = x_k + sp_k$
until $||\nabla F(x_{k+1})||$ is sufficiently small.

The vector p_k given by (6.1.6) is called the *Newton correction* and is based on regarding Q as a *local quadratic model* of F. Under favourable conditions – i.e. when the Hessian of F is positive definite – the Newton algorithm can be very efficient. The "natural" steplength implied by the quadratic model is $s = 1$; and in practice this often satisfies the Wolfe conditions and effectively eliminates the line search.

A worked example

We can demonstrate a typical Newton iteration using the function

$$F(x) = x_1^4 + 2x_2^2 + x_1 - x_2.$$

We shall avoid subscripts and let $x = (-\frac{1}{2}, \frac{1}{2})^T$ denote the starting point. We shall refer to the search direction as p and the new point obtained by the iteration will be written as $x^+ = x + sp$. Since

$$\nabla F = \begin{pmatrix} 4x_1^3 + 1 \\ 4x_2 - 1 \end{pmatrix} \quad \text{and} \quad \nabla^2 F = \begin{pmatrix} 12x_1^2 & 0 \\ 0 & 4 \end{pmatrix}$$

the Newton correction p is obtained by solving

$$3p_1 = -\frac{1}{2} \tag{6.1.7}$$

$$4p_2 = -1. \tag{6.1.8}$$

This gives $p_1 = -\frac{1}{6}$ and $p_2 = -\frac{1}{4}$ and so the new point is of the form

$$x^+ = (-\frac{1}{2} - \frac{s}{6}, \frac{1}{2} - \frac{s}{4})^T.$$

Using the "natural" steplength $s = 1$ we get $x^+ = (-\frac{4}{6}, \frac{1}{4})^T$ and the new value of F is

$$\frac{256}{1296} + \frac{2}{16} - \frac{4}{6} - \frac{1}{4} \approx -0.5942.$$

Since the value of F at the initial point is -0.4375, it is clear that the unit step, $s = 1$, has produced an acceptable reduction in F.

Exercises

1. In the worked example above, calculate the optimal step s^* for a perfect line search along the Newton direction. What are the largest values of η_1 and η_2 for which the step $s = 1$ satisfies the second and third Wolfe conditions?

2. Do one iteration of the Newton method applied to the function

$$F(x) = (x_1 - 1)^2 + x_2^3 - x_1 x_2$$

starting from $x_1 = x_2 = 1$. What happens when you start from $x = (1, -1)$?

2. Positive definiteness and Cholesky factors

In the Newton algorithm we must determine whether the Hessian G_k is positive definite. Very conveniently, the test for positive definiteness can be combined with the solution of $G_k p_k = -g_k$ if we use the method of *Cholesky factorization*. This seeks triangular factors of G_k so that

$$G_k = LL^T \qquad (6.2.1)$$

where L is a lower triangular matrix. Once we have found these factors we can solve $G_k p_k = -g_k$ by first obtaining the intermediate vector z to satisfy

$$Lz = -g_k \qquad (6.2.2)$$

and then getting p_k from

$$L^T p_k = z. \qquad (6.2.3)$$

The two linear systems (6.2.2) and (6.2.3) are easy to solve because they involve triangular coefficient matrices and so z and p_k are obtained by simple forward and backward substitution.

The Cholesky factorization (6.2.1) always exists if G_k is positive definite. Conversely, if G_k is not positive definite then the factorization process will break down. Attempting to find the Cholesky factors is usually the most efficient way of testing a symmetric matrix for positive definiteness.

We now describe the Cholesky method for solving a symmetric positive-definite linear system $Ax = b$ using the factorization $A = LL^T$. The method of calculating L is based on the fact that its elements must satisfy

$$A = \begin{pmatrix} l_{11} & 0 & \dots & 0 & \dots & 0 \\ l_{21} & l_{22} & \dots & 0 & \dots & 0 \\ & & \dots & \dots & & \\ l_{k1} & l_{k2} & \dots & l_{kk} & \dots & 0 \\ & & \dots & \dots & & \\ l_{n1} & l_{n2} & \dots & l_{nk} & \dots & l_{nn} \end{pmatrix} \begin{pmatrix} l_{11} & l_{21} & \dots & l_{k1} & \dots & l_{n1} \\ 0 & l_{22} & \dots & l_{k2} & \dots & l_{n2} \\ & & \dots & \dots & & \\ 0 & 0 & \dots & l_{kk} & \dots & l_{nk} \\ & & \dots & \dots & & \\ 0 & 0 & \dots & 0 & \dots & l_{nn} \end{pmatrix}$$

Considering just the first row of A, the rules of matrix multiplication imply

$$a_{11} = l_{11}^2; \quad a_{1j} = l_{11} l_{j1} \text{ for } j = 2,...,n.$$

Hence the first column of L can be obtained from

$$l_{11} = \sqrt{a_{11}}; \quad l_{j1} = \frac{a_{1j}}{l_{11}} \text{ for } j = 2,...,n.$$

In the second row of A we have

$$a_{22} = l_{21}^2 + l_{22}^2; \quad a_{2j} = l_{21} l_{j1} + l_{22} l_{j2} \quad \text{for } j = 3,..,n$$

and so the second column of L is given by

$$l_{22} = \sqrt{(a_{22} - l_{21}^2)}; \quad l_{j2} = \frac{(a_{j2} - l_{21} l_{j1})}{l_{22}} \quad \text{for } j = 3,...,n.$$

More generally, by considering the k-th row of A, we obtain the following expressions for the k-th column of L

$$l_{kk} = \sqrt{(a_{kk} - \sum_{i=1}^{k-1} l_{ki}^2)}; \quad l_{jk} = \frac{(a_{kj} - \sum_{i=1}^{k-1} l_{ki} l_{ji})}{l_{kk}} \quad \text{for } j = k+1,...,n. \quad (6.2.4)$$

A complete Cholesky factorization consists of applying (6.2.4) for $k = 1,...,n$. The process breaks down at stage k if the calculation of l_{kk} involves the square root of a negative number. This will not happen if A is positive definite.

As an example, consider the equations

$$\begin{array}{ccccccc} 10x_1 & + & x_2 & + & x_3 & = & 10 \\ x_1 & + & 8x_2 & + & 2x_3 & = & 7 \\ x_1 & + & 2x_2 & + & 20x_3 & = & -17 \end{array} \quad (6.2.5)$$

whose coefficient matrix is

$$A = \begin{pmatrix} 10 & 1 & 1 \\ 1 & 8 & 2 \\ 1 & 2 & 20 \end{pmatrix}.$$

The factorization process (6.2.4) gives

$$l_{11} = \sqrt{10} \approx 3.162; \quad l_{21} = l_{31} \approx \frac{1.0}{3.162} \approx 0.3163;$$

$$l_{22} \approx \sqrt{(8.0 - 0.3163^2)} \approx 2.811; \quad l_{32} \approx \frac{(2.0 - 0.3163 \times 0.3163)}{2.811} \approx 0.6759;$$

$$l_{33} \approx \sqrt{(20.0 - 0.3163^2 - 0.6759^2)} \approx \sqrt{19.44} \approx 4.409.$$

As a check we can show that the product LL^T gives a result which differs only slightly from A because of rounding errors in the four-digit arithmetic we used to compute the elements of L.

To complete the solution of (6.2.5) we deal first with the lower triangular system $Lz = b$, which is

$$\begin{pmatrix} 3.162 & 0 & 0 \\ 0.3163 & 2.811 & 0 \\ 0.3163 & 0.6759 & 4.409 \end{pmatrix} \begin{pmatrix} z_1 \\ z_2 \\ z_3 \end{pmatrix} = \begin{pmatrix} 10 \\ 7 \\ -17 \end{pmatrix}.$$

Forward substitution gives

$$z_1 \approx \frac{10}{3.162} \approx 3.163, \; z_2 \approx \frac{(7 - 0.3163 z_1)}{2.811} \approx 2.134$$

$$\text{and} \quad z_3 \approx \frac{(-17 - 0.6759 z_2 - 0.3163 z_1)}{4.409} \approx 4.409.$$

The upper triangular system $L^T x = z$ is

$$\begin{pmatrix} 3.162 & 0.3163 & 0.3163 \\ 0 & 2.811 & 0.6759 \\ 0 & 0 & 4.409 \end{pmatrix} \begin{pmatrix} x_1 \\ x_2 \\ x_3 \end{pmatrix} = \begin{pmatrix} 3.163 \\ 2.134 \\ 4.409 \end{pmatrix}.$$

and backward substitution then gives $x \approx (1.0, 1.0, -1.0)^T$. Clearly this satisfies the original system (6.2.5).

Exercises

1. Show that solving (6.2.2) and (6.2.3) yields the solution to $G_k p_k = -g_k$.

2. Solve the system of equations

$$\begin{array}{rcrcrcr} 10x_1 & + & x_2 & + & x_3 & = & 9 \\ x_1 & + & 8x_2 & + & 2x_3 & = & 11 \\ x_1 & + & 2x_2 & + & 12x_3 & = & -31 \end{array} \qquad (6.2.6)$$

using the Cholesky method.

3. Advantages & drawbacks of Newton's method

If the eigenvalues of the Hessian matrix $\nabla^2 F$ are bounded away from zero then it can be shown that the Newton iteration produces search directions which satisfy the first Wolfe condition (5.5.1). Therefore the Newton method converges if it uses a line search to satisfy the second and third Wolfe conditions. The rate of convergence can be quadratic, as stated in the following result.

Proposition If $F(x)$ is a function for which the Newton algorithm converges to a local minimum x^* and if the smallest eigenvalue of $\nabla^2 F(x^*)$ is $m > 0$ and if the third derivatives of $F(x)$ are bounded in some neighbourhood of x^* then there exists an integer $\bar{k}$ and a positive real constant $K(< 1)$ such that, for $k > \bar{k}$,

$$||x_{k+1} - x^*|| < K||x_k - x^*||^2.$$

(The essentials of the proof are similar to the one variable case in chapter 2.)

It is important to point out that such theoretical convergence rates are not always observed in practice because of *rounding errors* in computer arithmetic. Since the results of all calculations must be expressed in some fixed number of digits (typically about 14 in double precision arithmetic) there will inevitably be small errors in computed values of F, and the elements of ∇F and $\nabla^2 F$ during the solution of an optimization problem. These errors are usually negligible: but they can become significant when ∇F is near zero. This can prevent iterative methods from reaching solutions with arbitrarily high accuracy. (For a fuller account of rounding errors see the text by Higham [12].)

In spite of the above cautionary remarks, however, the theoretical quadratic convergence of the Newton method does imply that it can be very efficient. Unfortunately, however, the method also has some drawbacks.

(i) The process of calculating all the required second derivatives can be laborious, especially if the differentiation is done by hand. This – and the subsequent coding of the derivative expressions – is both time consuming and liable to error. It is possible to approximate second derivatives by finite difference expressions such as

$$\frac{\partial^2 F(x)}{\partial x_i \partial x_j} \approx \frac{1}{h}\left[\frac{\partial F(x + he_j)}{\partial x_i} - \frac{\partial F(x)}{\partial x_i}\right];$$

but these can be inaccurate if h is badly chosen. *Automatic differentiation* techniques [13, 14] can now simplify the task of obtaining second derivatives: but, in the past, the Newton method has often been neglected because of the amount of work it entails.

(ii) The Newton method is expensive because it solves a system of linear equations to obtain the search direction. The Cholesky method is more efficient

than the general-purpose Gaussian elimination method but it still uses $O(\frac{1}{6}n^3)$ multiplications per iteration.

(iii) As we have already mentioned, the Cholesky solution of $G_k p_k = -g_k$ may break down because G_k is not positive definite. This leads to the most serious difficulty for the Newton approach – namely that it does not invariably provide a good search direction. As we observed in the one variable case, the Newton method may be attracted to a local maximum or saddle point in regions where the Hessian is not positive definite. The algorithm given in section 1 deals with the possibility of unsuitable search directions by resorting to the steepest descent direction on certain iterations. Better strategies than this can be devised; but the fact still remains that the Newton algorithm, in practice, does require a "fall-back option" to ensure that it will converge. Further discussion of this appears in the next section.

Exercise
If a search direction is obtained by solving an $n \times n$ linear system $Bp = -g$ and if B is positive definite show that p is a descent direction. If the eigenvalues of B are bounded above by M and below by m show that Wolfe condition 1 is satisfied with $\eta_0 = \sqrt{(m/M)}$.

4. Search directions from indefinite Hessians

Matrix modification techniques

Suppose that, during a Newton iteration, the Cholesky factorization breaks down at step k because the calculation of l_{kk} in (6.2.4) involves the square root of a negative argument. One way of trying to continue with the calculation of a search direction would be to modify the Hessian matrix by increasing the k-th diagonal term a_{kk} so that it exceeds $\sum_{i=1}^{k-1} l_{ki}^2$. Hence we could change the true Hessian *during the factorization* in order to obtain factors L, L^T of a matrix $\hat{G}_k$ which differs from G_k in one (or more) diagonal elements. These triangular factors can then be used to solve $\hat{G}_k p = -g_k$. This will give p as a descent direction, based on *partial* second derivative information, which can be used as a substitute for the Newton direction

In practice, the modification of G_k by the method outlined in the previous paragraph does not always work very well because the computed L and L^T factors may contain very large elements. Gill & Murray [15] and Schnabel & Eskow [16] have suggested more complicated – but more numerical stable – ways of changing the Hessian during the Cholesky process so as to get L and L^T as factors of a matrix $\hat{G}_k$ which may differ from G_k in both diagonal and off-diagonal terms.

The SAMPO procedure which implements the Newton method uses an alternative fall-back strategy whenever the Cholesky factorization fails. The Gershgorin disk theorem [17] implies that a symmetric matrix A will have all positive eigenvalues if it satisfies a diagonal dominance condition

$$a_{kk} > \sum_{j=1,\ j\neq k}^{n} |a_{kj}|.$$

Hence, if the Cholesky factorization breaks down we can obtain a modified Hessian $\hat{G}$ by increasing the diagonal elements of G, where necessary, so that

$$g_{kk} \geq 1.1 \sum_{j=1,\ j\neq k}^{n} |g_{kj}|.$$

Trust-region methods

If G_k is not positive definite we can get a downhill search direction by solving

$$(\lambda I + G_k)p_k = -g_k \tag{6.4.1}$$

for a "suitably large" value of the positive scalar λ. This is because the eigenvalues of $(\lambda I + G_k)$ exceed those of G_k by λ and hence, when λ is big enough, $(\lambda I + G_k)$ must be positive definite.

Using a search direction given by (6.4.1) might seem as arbitrary as the matrix modification approach in section 4, above. However, it turns out that (6.4.1) gives p_k as a solution of a sub-problem of the form

$$\text{Minimize} \quad Q(p) = \frac{1}{2}p^T G_k p + p^T g_k \quad \text{subject to} \quad ||p||_2 \leq \Delta. \tag{6.4.2}$$

In other words, $x_k + p_k$ minimizes a quadratic approximation to F subject to an upper bound on the size of the move away from x_k. (The connection between (6.4.2) and (6.4.1) will be established in a later chapter.)

Problem (6.4.2) always has a solution even when G_k is not positive definite. This is because it simply amounts to finding the smallest value of $Q(p)$ within a hyperspherical region around x_k. Hence (6.4.2) provides a reasonable way of getting round one of the main drawbacks of the Newton method.

Problem (6.4.2) is the basis for a class of minimization techniques known as *Trust Region methods* fully described by Conn, Gould and Toint [18]. All the methods we have considered so far have employed the strategy of choosing a promising search direction first and then determining a step-size by a line search. We could reverse this approach and decide first on a suitable step-size and then work out a direction in which to take it.

Suppose for instance that we have some grounds for believing we can *trust* a local quadratic model of F to be reasonably accurate providing we stay within a distance $||\Delta||$ of the current iterate x_k. A new point $x_{k+1} = x_k + p_k$ could then be obtained by solving (6.4.2) *whether G_k is positive definite or not*. The *trust region radius*, Δ, can be adjusted from iteration to iteration. It is increased if the actual change $F(x_{k+1}) - F(x_k)$ agrees well with the predicted change in the quadratic model. Conversly, it is decreased if the actual and predicted changes are too inconsistent. Ultimately, the value of Δ becomes large enough for the subproblem (6.4.2) to allow full Newton steps to be taken and hence permit quadratic convergence.

One disadvantage of the trust-region approach is that (6.4.2) can be difficult and expensive to solve accurately on each iteration. The relationship between the trust region radius Δ and the value of λ in (6.4.1) is highly non-linear and it is not usually possible to obtain p_k to solve (6.4.2) via a single solution of (6.4.1). Therefore most implementations make do with an approximate solution. A possible approach could involve reducing (6.4.2) to a 2-D problem. For instance we could combine the negative gradient $-g_k$ with the Newton direction p_k (even if this is uphill) [19]. This would mean seeking a new point $x_{k+1} = x_k - \alpha g_k + \beta p_k$ that gives the least value of F in the plane $(p_k, -g_k)$, subject to a restriction on step-size. Better still, for the non-positive-definite case, would be to determine a direction of *negative curvature* – i.e., a vector v such that $v^T G_k v < 0$ (see [20]). We could then search for the minimum of F in a plane defined by $-g_k$ and v.

Exercise
If $\bar{G} = \lambda I + G_k$ show that $x^T \bar{G} x > 0$ for all $x \neq 0$, when λ is sufficiently large.

5. Numerical results with the Newton method

We use NMp and NMw, respectively, to denote the SAMPO implementations of the Newton method with perfect and weak line searches. They both use the matrix modification strategy in section 4 to generate descent directions when the Hessian is not positive-definite. Table 6.1 shows numbers of iterations and function calls needed to solve problems T1 – T4 to standard accuracy (4.4.2). (Results in this section were obtained with program `sample3`.)

	T1 $\rho = 100$	T2 $\rho = 50$	T3 $\rho = 2$	T4 $\rho = 50$
NMp	1/2	1/2	12/68	8/51
NMw	1/2	1/2	20/29	16/19

Table 6.1. Performance of NMp and NMw on problems T1 – T4

Some conclusions to be drawn from Table 6.1 are as follows.
• Both NMp and NMw converge in just one iteration on problems T1 and T2. This is because **Minrisk1m** has a quadratic objective function and so a single Newton step (6.1.6) locates the minimum.
• On all the test problems, the Newton method converges in considerably fewer iterations and function calls than steepest descent (see Table 5.3).
• On the non-quadratic problems, NMw is typically more economical than NMp in terms of function evaluations even when it requires more iterations.

The Newton method has to deal with non-positive-definite Hessian matrices during some of the solutions quoted in Table 6.1. On Problem T4, for instance, NMp encounters an indefinite Hessian on iteration three. However, the modified Cholesky factorization generates a suitable descent direction and the method goes on to converge to the correct solution. When NMw is applied to the same problem it takes different steps on the first two iterations and (by chance rather than design) this causes the search to avoid the nonconvex region and all the iterations use the unmodified Newton direction.

Table 6.2 shows the work done by SDw and NMw to obtain both low- and standard-accuracy solutions of problem T4 (with $\rho = 50$). This demonstrates the practical implications of the Newton method's theoretical quadratic convergence rate. The linearly convergent steepest descent is obviously making *very* slow progress near the solution and requires over 700 iterations to reduce $||\nabla F||$ from $O(10^{-4})$ to $O(10^{-5})$. Newton's method, on the other hand, takes only one iteration to yield a similar improvement.

	itns/fns (low)	itns/fns (standard)
SDw	315/934	1047/3130
NMw	15/18	16/19

Table 6.2. Ultimate convergence of SDw and NMw on problem T4

Exercises (To be solved using sample3 or other suitable software.)
1. For the data in Table 1.2, use Newton's method to solve the maximum return problem with target risk $V_a = 0.00075$ using $\rho = 0.1$. How does the solution compare with what you get when $\rho = 1$ and 10? Explain the differences.

2. For the dataset Real-20-5 (with $\bar{r}$ and Q given by (4.4.7), (4.4.8)) use Newton's method to determine the range of values for acceptable risk V_a for which the maximum-return solution does *not* involve short-selling.

3. Compare the performance of the Newton method and steepest descent when applied to problem T3 but using larger values of ρ in the range $10 \le \rho \le 100$.

4. Suppose we add a *riskless asset* to the data in Table 4.1 – that is, we also include a sixth asset whose return is a constant 1%. How does the solution of problem T1 change? What happens if a further riskless asset (with $r_7 = 0.5\%$) is added to the portfolio?

5. Use the transformation $y_i = x_i$, $i = 1,..,n-1$, $y_n = 1 - \sum_{i=1}^{n} x_i$ to rewrite

$$F = \frac{V}{R} = \frac{y^T Q y}{\bar{r}^T y}$$

as a function of $x_1,...,x_{n-1}$. Hence determine expressions for $\partial F / \partial x_i$ and $\partial^2 F / \partial x_i \partial x_j$ and apply Newton's method to minimize F when $\bar{r}$ and Q are given by (4.4.7) and (4.4.8).

6. Use the same transformation as in question 5 to express the function

$$F = \frac{(y^T Q y - V_a)^2}{\bar{r}^T y}$$

in terms of independent variables $x_1,...,x_{n-1}$. Hence obtain the gradient and Hessian with respect to x and use Newton's method to minimize the function for the data in Table 1.1 and with $V_a = 0.003$. How does your result compare with the solution of problem T3?

Steepest descent

Still the fog persists.
Let the incline have its way
and set your compass.

Keep taking footsteps
until that first suspicion
of an uphill slope

then turn left or right,
just one of many zig-zags.
Will it ever end?

Chapter 7

QUASI-NEWTON METHODS

1. Approximate second derivative information

Some drawbacks of the Newton method were mentioned in the previous chapter. These have motivated the development of *Quasi-Newton* (or *Variable-metric*) techniques. The essential idea of these methods is simple. A positive definite matrix is used to *approximate* the Hessian matrix (or its inverse). This saves the work of computing exact second derivatives and also avoids the difficulties associated with loss of positive definiteness. The approximate matrix is *updated* on each iteration so that, as the search proceeds, second derivative information is improved. Before going into detail about this updating we give an outline quasi-Newton algorithm.

An outline quasi-Newton Method

Choose x_0 as an initial estimate of the minimum of $F(x)$.
Choose H_0 as an arbitrary symmetric positive definite matrix
Repeat for $k = 0, 1, 2, ...$
Set $g_k = \nabla F(x_k)$
Set $p_k = -H_k g_k$
Find s, so $F(x_k + sp_k)$ satisfies (5.5.2), (5.5.3) for some η_1, η_2
set $x_{k+1} = x_k + sp_k$, $\gamma_k = g_{k+1} - g_k$, $\delta_k = x_{k+1} - x_k$
Obtain a new positive definite matrix H_{k+1} such that

$$H_{k+1}\gamma_k = \delta_k \tag{7.1.1}$$

until $||\nabla F(x_{k+1})||$ is sufficiently small.

In this algorithm, H_k is an estimate of the inverse Hessian $\nabla^2 F(x_k)^{-1}$. The simple initial choice $H_0 = I$, the identity matrix, is usually satisfactory.

Definition The equation (7.1.1) used in the calculation of the new matrix H_{k+1} is called the *quasi-Newton condition.*

Condition (7.1.1) is derived as follows. If $F(x) = \frac{1}{2}x^T Ax + b^T x + c$ then

$$\gamma_k = g_{k+1} - g_k = (Ax_{k+1} + b) - (Ax_k + b) = A(x_{k+1} - x_k) = A\delta_k.$$

In other words

$$A^{-1}\gamma_k = \delta_k.$$

Thus, when $F(x)$ is a quadratic function, the condition (7.1.1) causes H_{k+1} to share a property with the true inverse Hessian.

To save computing effort – and also to preserve second-derivative information already present in H_k – the new matrix H_{k+1} is obtained by a *low-rank* modification to H_k. We shall now describe some well-known *updating formulae.*

2. Rank-two updates for the inverse Hessian

Definition The *Davidon-Fletcher-Powell* (*DFP*) update [21, 22] for H_{k+1} is

$$H_{k+1} = H_k - \frac{H_k\gamma_k\gamma_k^T H_k}{\gamma_k^T H_k\gamma_k} + \frac{\delta_k\delta_k^T}{\delta_k^T\gamma_k}. \tag{7.2.1}$$

Proposition The *DFP* formula makes H_{k+1} satisfy (7.1.1).
Proof The result follows immediately on multiplying the right hand side of (7.2.1) by γ_k and simplifying.

Proposition The *DFP* formula causes H_{k+1} to inherit positive definiteness from H_k provided

$$\delta_k^T\gamma_k > 0. \tag{7.2.2}$$

(The proof of this is left to the reader – see Exercise 3, below.)

The quasi-Newton condition does not define H_{k+1} uniquely since it consists of n equations involving the n^2 elements of H_{k+1}. As well as the *DFP* formula, the *Broyden-Fletcher-Goldfarb-Shanno* (*BFGS*) formula [23, 24] also causes H_{k+1} to satisfy (7.1.1).

Definition The *BFGS* formula for H_{k+1} is

$$H_{k+1} = H_k - \frac{H_k\gamma_k\delta_k^T + \delta_k\gamma_k^T H_k}{\delta_k^T\gamma_k} + [1 + \frac{\gamma_k^T H_k\gamma_k}{\delta_k^T\gamma_k}]\frac{\delta_k\delta_k^T}{\delta_k^T\gamma_k}. \tag{7.2.3}$$

This formula also ensures that H_{k+1} is positive definite when (7.2.2) holds.

An important result which links the *DFP* and *BFGS* updates is the following.
Dixon's Theorem [25, 26] If a quasi-Newton algorithm includes a perfect line search then, for any function $F(x)$, the same sequence of iterates $\{x_k\}$ will be obtained irrespective of whether H_k is produced by the *DFP* or *BFGS* formula.

This theorem seems to imply there is no practical difference between the *DFP* and *BFGS* updates. However, when we attempt a perfect line search in finite precision arithmetic, rounding errors can prevent the termination condition $p_k^T g_{k+1} = 0$ from being satisfied precisely. It turns out that even small departures from "perfection" in the line search can cause differences to appear in the iterates given by different updates. Moreover, most quasi-Newton implementations use weak line searches and then Dixon's theorem does not apply.

Experience suggests that, in practice, the *BFGS* update is usually preferable to the *DFP*. It appears that, while both (7.2.1) and (7.2.3) keep H_k positive definite, the *DFP* formula is more likely to produce matrices which are near-singular and this can have an adverse affect on its performance.

A worked example

We consider a quasi-Newton iteration (with perfect line search and *DFP* update) on the function

$$F(x) = x_1^2 + 3x_2^2 + x_1x_2 + x_1 + x_2.$$

We dispense with iteration-number subscripts for this example and take the starting point as $x = (0,\ 0)^T$. To make the iteration differ from steepest descent we use the initial inverse Hessian estimate

$$H = \begin{pmatrix} \frac{1}{2} & 0 \\ 0 & \frac{1}{6} \end{pmatrix}.$$

The gradient of $F(x)$ is

$$g = \begin{pmatrix} 2x_1 + x_2 + 1 \\ 6x_2 + x_1 + 1 \end{pmatrix}$$

and therefore the initial search direction is

$$p = -Hg = -\begin{pmatrix} \frac{1}{2} & 0 \\ 0 & \frac{1}{6} \end{pmatrix} \begin{pmatrix} 1 \\ 1 \end{pmatrix} = \begin{pmatrix} -\frac{1}{2} \\ -\frac{1}{6} \end{pmatrix}.$$

The new point is $x^+ = x + sp = (-\frac{s}{2},\ -\frac{s}{6})^T$ where s is chosen to minimize

$$F(x+sp) = \frac{s^2}{4} + \frac{3s^2}{36} + \frac{s^2}{12} - \frac{s}{2} - \frac{s}{6} = \frac{5s^2}{12} - \frac{2s}{3}.$$

This gives $s^* = 0.8$ and so $x^+ = (-0.4, -0.1333)^T$. The new gradient is

$$g^+ = \begin{pmatrix} 0.0667 \\ -0.2 \end{pmatrix}$$

and, using quasi-Newton notation,

$$\delta = x^+ - x = \begin{pmatrix} -0.4 \\ -0.1333 \end{pmatrix}, \quad \gamma = g^+ - g = \begin{pmatrix} -0.9333 \\ -1.2 \end{pmatrix}.$$

Thus, working to five significant figures, $\delta^T\gamma = 0.53328$. Moreover,

$$H\gamma = \begin{pmatrix} -0.46665 \\ -0.2 \end{pmatrix}$$

and so $\gamma^T H\gamma = 0.67552$. We also obtain

$$H\gamma\gamma^T H = \begin{pmatrix} 0.21776 & 0.09333 \\ 0.09333 & 0.04 \end{pmatrix} \quad \text{so that} \quad \frac{H\gamma\gamma^T H}{\gamma H\gamma} = \begin{pmatrix} 0.32236 & 0.13816 \\ 0.13816 & 0.05921 \end{pmatrix}$$

and

$$\delta\delta^T = \begin{pmatrix} 0.16 & 0.05332 \\ 0.05332 & 0.017769 \end{pmatrix} \quad \text{so that} \quad \frac{\delta\delta^T}{\delta^T\gamma} = \begin{pmatrix} 0.3 & 0.1 \\ 0.1 & 0.03333 \end{pmatrix}.$$

Putting these ingredients together in the *DFP* formula

$$H^+ = \begin{pmatrix} 0.5 - 0.32236 + 0.3 & 0 - 0.13816 - 0.1 \\ 0 - 0.13816 - 0.1 & 0.16667 - 0.05921 + 0.03333 \end{pmatrix}$$

$$= \begin{pmatrix} 0.47764 & -0.03816 \\ -0.03816 & 0.14079 \end{pmatrix}.$$

On the next iteration, the search direction will be

$$p = -H^+g^+ = \begin{pmatrix} -0.039474 \\ 0.030702 \end{pmatrix}$$

and the reader can verify that the perfect step $s^* \approx 1.3807$ along p away from x^+ will locate the minimum of $F(x)$ at $(-\frac{5}{11}, -\frac{1}{11})^T$ (subject to rounding errors in five-digit arithmetic).

Exercises

1. For the worked example above, does the second iteration locate the solution if H^+ is obtained by the *BFGS* update?

2. Using the *DFP* update and perfect line searches, do two quasi-Newton iterations on the function

$$F(x) = x_1^2 + x_1 x_2 + \frac{x_2^2}{2}$$

starting from $x = (1, 1)$. What happens if a weak line search is used instead?

3. Prove that the *DFP* update ensures that H_{k+1} inherits positive definiteness from H_k provided $\delta_k^T \gamma_k > 0$. (**Hint**: If H_k is positive definite then it has a Cholesky factor L such that $H_k = LL^T$.)

4. Prove that the condition (7.2.2) for ensuring positive definiteness in *DFP* and *BFGS* updates is automatically satisfied when a perfect line search is used.

5. If F is a quadratic function and if $H_k = (\nabla^2 F(x_k))^{-1}$ show that H_{k+1} given by the *DFP* update is equal to $(\nabla^2 F(x_{k+1}))^{-1}$. Is the same true if H_{k+1} is given by the *BFGS* update?

5. Show that the following general result follows from Dixon's Theorem:
If a quasi-Newton algorithm includes a perfect line search then, for any function $F(x)$, the same sequence of iterates $\{x_k\}$ will be produced when the update for H_{k+1} is any member of the *family* defined by

$$H_{k+1} = \theta H_{k+1}^{dfp} + (1 - \theta) H_{k+1}^{bfgs} \tag{7.2.4}$$

where $1 \geq \theta \geq 0$ and H_{k+1}^{dfp}, H_{k+1}^{bfgs} denote the right hand sides of the updating formulae (7.2.1) and (7.2.3).

3. Convergence of quasi-Newton methods

There are a number of convergence results about quasi-Newton methods based on the *DFP* and *BFGS* updates. The following propositions all assume exact arithmetic is used.

Proposition [23] If $F(x)$ is an n-variable convex quadratic function then a quasi-Newton algorithm, with perfect line search, will converge to the minimum of F in at most n iterations with both the *DFP* and *BFGS* update. Moreover $H_n = \nabla^2 F^{-1}$.

Proposition (Powell [27]) If $F(x)$ is a twice-differentiable function which is convex in some region R around a local minimum x^*, then a quasi-Newton algorithm, with perfect line search and either the *DFP* or *BFGS* update, will converge to x^* from any starting point in R.

Proposition (Powell [28]) If $F(x)$ is a twice-differentiable function which is convex in some region R around a local minimum x^*, then a quasi-Newton

algorithm, with a weak line search and the *BFGS* update, will converge to x^* from any starting point in R.
(Note that a similar result about convergence of a quasi-Newton algorithm with a weak line search and the *DFP* update has also been proved [29] and it shows that stronger conditions on steplength are needed than for the *BFGS* version.)

Because they do not use the exact Hessian, quasi-Newton methods do not usually converge as quickly as the Newton method. Performance near the solution is, however, superior to that of the steepest descent approach.

Proposition [27] If H_k tends to the true inverse Hessian as x_k approaches x^* and if the step size $s = 1$ satisfies Wolfe conditions (5.5.2), (5.5.3) for all $k \geq K$ then quasi-Newton methods are capable of ultimately *superlinear* convergence. This means that, for k sufficiently large,

$$\frac{||x_{k+1} - x^*||}{||x_k - x^*||} \to 0$$

or, equivalently, that the error norm decreases at a rate implied by

$$||x_{k+1} - x^*|| = C||x_k - x^*||^r$$

for some constant C and for $1 < r < 2$. This is not as good as the quadratic $(r = 2)$ convergence given by the Newton method; but it is superior to the linear $(r = 1)$ convergence of the steepest descent algorithm.

Since the updating formulae for H_{k+1} involve only vector-vector and matrix-vector products, the number of multiplications per iteration of a quasi-Newton method varies with n^2. This compares favourably with the $O(n^3)$ multiplications per iteration needed by the Newton method to form and factorize the Hessian $\nabla^2 F(x)$. On the other hand, the Newton method may take significantly fewer iterations and so there is not always a clear-cut advantage in run-time for quasi-Newton methods.

4. Numerical results with quasi-Newton methods

The `SAMPO` implementations of the quasi-Newton approach are called `QNp` and `QNw` to denote the use of a perfect or a weak line search. In both cases the *BFGS* formula (7.2.3) is used to update the approximate inverse Hessian. The program `sample3` can solve **Minrisk1m** and **Maxret1m** by means of these quasi-Newton methods.

Table 7.1 shows numbers of iterations and function calls needed to solve problems T1 – T4. As in Tables 5.3 and 6.1, the convergence test is (4.4.2).

	T1 $\rho = 100$	T2 $\rho = 50$	T3 $\rho = 2$	T4 $\rho = 50$
QNp	4/13	4/10	22/119	22/120
QNw	9/15	6/11	33/48	26/51

Table 7.1. Performance of QNp and QNw on problems T1 – T4

Noteworthy points about Table 7.1 are as follows.

• Convergence of QNp on problems T1 and T2 matches theoretical expectations in the first proposition of section 3. Problems T1 and T2 both involve five assets and so **Minrisk1m** is a quadratic problem in four variables. Thus a quasi-Newton method with perfect linesearch should converge in (at most) four iterations. Note that, while QNw needs more iterations than QNp, it uses fewer function calls per linesearch.

• On the non-quadratic problems T3 and T4, QNw typically uses more iterations than QNp; but in terms of overall workload this is balanced by a decrease in function calls per iteration of QNw.

• Comparison with Tables 5.3 and 6.1 shows that the quasi-Newton approach is quite competitive with the Newton method and is considerably more efficient than steepest descent. (We shall no longer consider the steepest descent method as a serious contender for solving practical problems.)

Table 7.2 illustrates the differences in the ultimate convergence of Newton and quasi-Newton methods by showing numbers of iterations and function calls needed to solve problem T4 ($\rho = 50$) to both low- and high-accuracy (convergence tests (4.4.3)).

	itns/fns (low)	itns/fns (high)
NMw	15/18	16/19
NMp	8/55	9/56
QNw	26/51	35/60
QNp	21/118	24/124

Table 7.2. Ultimate convergence of NM and QN on problem T4

With both the perfect and the weak linesearch, the quadratically convergent Newton method only needs one iteration to go from low to high accuracy. Quasi-Newton methods are only capable of superlinear convergence and so QNw takes nine iterations to reduce the gradient norm from $O(10^{-4})$ to $O(10^{-6})$ while QNp needs three. This quasi-Newton performance is, however, much better than that of the linearly convergent steepest descent method in Table 6.2.

Exercises (To be solved using `sample3` or other suitable software.)
1. Use a quasi-Newton method to solve the minimum risk problem for data from Table 4.1. Use several target returns R_p in the range $0.75 \le R_p \le 1.5$ and plot the corresponding values of minimum risk. On each problem you solve, compare the numbers of iterations and function calls needed when perfect and weak linesearches are used.

2. Use a quasi-Newton method to solve the maximum return problem for the data in Table 4.1. Take values of V_a in the range $0.002 \le V_a \le 0.008$ (setting $\rho = 10$ in all cases) and plot the resulting maximum returns against V_a.

3. Suppose that R_a is the maximum return obtained by solving **Maxret1m** for a particular value of V_a. If **Minrisk1m** is solved with $R_p = R_a$, does it give a solution with $V \approx V_a$? Is the converse result true?

4. An alternative approach to the maximum-return problem would be to solve

$$\text{Minimize} \quad F(x) = \frac{1}{\bar{r}^T y} + \frac{\rho}{V_a^2}(y^T Q y - V_a)^2 \quad \text{subject to} \quad \sum_{i=1}^{n} y_i = 1. \quad (7.4.1)$$

instead of **Maxret1m.** Use the transformation based on (3.1.5) to express (7.4.1) as a function of x, where $x_i = y_i, i = 1,..,n-1$.
Minimize this function by both the steepest descent and the quasi-Newton methods, using data from Table 1.1 and setting $\rho = 10$ and $V_a = 0.003$. Does this seem to be a better formulation than **Maxret1m**?

5. Implement a new quasi-Newton which uses the *DFP* update instead of the *BFGS* formula. Test its performance on the problems suggested elsewhere in these exercises.

5. The rank-one update for the inverse Hessian

Definition The *Symmetric Rank One* (*SR1*) updating formula is

$$H_{k+1} = H_k + \frac{v_k v_k^T}{v_k^T \gamma_k} \quad \text{where } v_k = (\delta_k - H_k \gamma_k). \qquad (7.5.1)$$

The *SR1* update differs from the *DFP* and *BFGS* formulae which change the matrix H_k in the *two*-dimensional subspace spanned by the vectors δ_k and $H_k\gamma_k$. Formula (7.5.1) only alters H_k in the space spanned by the vector $(\delta_k - H_k\gamma_k)$ and has an interesting "memory property" when used with quadratic functions.

Proposition If $H_k\gamma_{k-1} = \delta_{k-1}$ (so that H_k agrees with the true inverse Hessian for the vectors γ_{k-1} and δ_{k-1}) then the matrix H_{k+1} given by (7.5.1) satisfies

$$H_{k+1}\gamma_k = \delta_k \quad \text{and} \quad H_{k+1}\gamma_{k-1} = \delta_{k-1}.$$

Proof of this property is left to the reader. From it there follows

Proposition If $F(x)$ is an n-variable convex quadratic function then a quasi-Newton algorithm, using a weak line search and the *SR1* update will obtain $H_n = \nabla^2 F^{-1}$, and therefore will converge to the minimum of F in at most $n+1$ iterations.

In one sense, *SR1* is better than *DFP* or *BFGS* since it gives finite termination on a quadratic function without perfect line searches. However (7.5.1) has the drawback that it may not keep H_{k+1} positive definite under the same, easy to check, condition (7.2.2) as applies to the *DFP* or *BFGS* formulae. Indeed even when *SR1* is used on a positive definite quadratic function some of the intermediate H_k may be indefinite.

Exercises
1. Prove the "memory property" of the symmetric rank-one update, namely: If $H_k \gamma_{k-1} = \delta_{k-1}$ then the matrix H_{k+1} given by (7.5.1) satisfies

$$H_{k+1}\gamma_k = \delta_k \quad \text{and} \quad H_{k+1}\gamma_{k-1} = \delta_{k-1}.$$

2. Do two quasi-Newton iterations with weak search and *SR1* update on

$$F(x) = x_1^2 + x_1 x_2 + \frac{x_2^2}{2}$$

starting from $x = (1,1)$. Comment on the outcome.

3. Derive a condition which will ensure that H_{k+1} given by the *SR1* update will inherit positive definiteness from H_k.

4. Implement a quasi-Newton procedure which uses the *SR1* update. How does it perform on problems suggested elsewhere in this section? (Your answer should deal with both perfect and weak line searches.)

6. Updating estimates of the Hessian

Some implementations of the quasi-Newton technique work with estimates, B_k, of $\nabla^2 F$ rather than $\nabla^2 F^{-1}$. (It can be argued that approximating the Hessian is a more numerically stable process than approximating its inverse.) The quasi-Newton condition for B_{k+1} is, of course,

$$B_{k+1}\delta_k = \gamma_k. \tag{7.6.1}$$

It can be shown that the *DFP* and *BFGS* formulae are *dual* in the following sense. If $H_k = (B_k)^{-1}$ then the update which gives $B_{k+1} = (H_{k+1}^{bfgs})^{-1}$ is

$$B_{k+1} = B_k - \frac{B_k\delta_k\delta_k^T B_k}{\delta_k^T B_k\delta_k} + \frac{\gamma_k\gamma_k^T}{\delta_k^T\gamma_k}. \qquad (7.6.2)$$

This is precisely the *DFP* formula with B replacing H and with δ and γ interchanged. Similarly $B_{k+1} = (H_{k+1}^{dfp})^{-1}$ is found by replacing H with B and exchanging δ and γ in the *BFGS* update.

The *SR1* formula (7.5.1) is *self-dual*, i.e. $B_{k+1} = (H_{k+1}^{sr1})^{-1}$ is given by

$$B_{k+1} = B_k + \frac{w_k w_k^T}{w_k^T\delta_k} \quad \text{where } w_k = (\gamma_k - B_k\delta_k). \qquad (7.6.3)$$

It might seem inefficient in practice to use an algorithm involving B instead of H since p_k will then be obtained from

$$B_k p_k = -g_k \qquad (7.6.4)$$

which implies that B_k must be factorized. This costly process is avoided when H_k approximates the inverse Hessian. However, Gill & Murray [15] have shown that it is possible to store and update the Cholesky factors of B_k. This makes it much more economical to solve (7.6.4) on every iteration.

Much more work has been done on the theory and implementation of quasi-newton methods than can be contained in a single chapter. For fuller accounts of other updating formulae and algorithms see [13] and [30].

Exercises

1. If a non-singular matrix Q is updated to become $\tilde{Q} = Q + uu^T$ show that

$$\tilde{Q}^{-1} = Q^{-1} - \frac{Q^{-1}uu^TQ^{-1}}{1+u^TQ^{-1}u}. \qquad (7.6.5)$$

(This is called the *Sherman-Morrison-Woodbury* formula.)

2. Use (7.6.5) to show that the SR1 formula is self-dual – i.e. that if $H_k = B_k^{-1}$ and if B_{k+1} is given by (7.6.3) then $H_{k+1}^{sr1} = B_{k+1}^{-1}$.

Chapter 8

CONJUGATE GRADIENT METHODS

1. Conjugate gradients and quadratic functions

We have already shown that the positive definite quadratic function

$$Q(x) = \frac{1}{2}(x^T Ax) + b^T x + c$$

can be minimized by solving $Ax = -b$. This can be done by the *conjugate gradient method* [31] outlined below, in which g denotes $\nabla Q = Ax + b$.

Conjugate gradient method for solving $Ax = -b$

Choose x_0 as an initial estimate of the solution
Calculate $g_0 = Ax_0 + b$. Set $p_0 = -g_0$
Repeat for $k = 0, 1, 2, ...$
find s so that $p_k^T g_{k+1} = p_k^T (A(x_k + sp_k) + b) = 0$
set $x_{k+1} = x_k + sp_k$
set β using

$$\beta = \frac{g_{k+1}^T g_{k+1}}{g_k^T g_k} \tag{8.1.1}$$

set $p_{k+1} = -g_{k+1} + \beta p_k$
until $||g_{k+1}|$ is sufficiently small.

The step, s, along the search direction p_k in the algorithm is given by

$$s = -\frac{p_k^T g_k}{p_k^T A p_k}. \tag{8.1.2}$$

This gives $p_k^T g_{k+1} = 0$, is equivalent to choosing s to minimize $Q(x_k + sp_k)$.

The formula for β used in the recurrence relation (8.1.1) is designed to make the search directions *conjugate* with respect to A.

Definition Two vectors u and v are said to be conjugate with respect to a symmetric matrix A if

$$u^T A v = 0. \tag{8.1.3}$$

Hence the search directions in the conjugate gradient algorithm satisfy

$$p_i^T A p_j = 0 \quad \text{when } i \neq j. \tag{8.1.4}$$

To show the significance of making the search directions mutually conjugate with respect to A, we consider the first two iterations of the conjugate gradient algorithm. At the end of iteration one the new point is x_1 and the gradient $g_1 = Ax_1 + b$. Because of the perfect line search we also have $p_0^T g_1 = 0$. Now consider iteration two. It will generate a point

$$x_2 = x_1 + sp_1 \quad \text{where} \quad g_2 = Ax_2 + b \quad \text{and} \quad p_1^T g_2 = 0.$$

We can now show that g_2 also satisfies

$$p_0^T g_2 = 0. \tag{8.1.5}$$

This means that the gradient after two iterations is orthogonal to *both* search directions p_0 and p_1. To prove (8.1.5) we note that

$$p_0^T g_2 = p_0^T(Ax_1 + sAp_1 + b) = p_0^T g_1 + sp_0^T A p_1.$$

The first term in the rightmost expression is zero because of the line search on iteration one. The second is zero because p_0 and p_1 are conjugate w.r.t. A. Hence (8.1.5) holds.

By considering an extension of (8.1.5) for k (> 2) iterations (Exercise 3 below) it is possible to prove the following important *finite termination property*.

Proposition In exact arithmetic, the conjugate gradient method solves an $n \times n$ system $Ax = -b$ in at most n iterations.

A worked example

We apply the conjugate gradient method to the function

$$f(x) = x_1^2 + x_1 x_2 + \frac{x_2^2}{2}$$

starting from $x_0 = (1,1)^T$. The gradient vector is

$$g = \begin{pmatrix} 2x_1 + x_2 \\ x_1 + x_2 \end{pmatrix}$$

and so the search direction away from x_0 is

$$p_0 = -g_0 = \begin{pmatrix} -3 \\ -2 \end{pmatrix}.$$

Hence the new point will be of the form $x_1 = (1 - 3s,\ 1 - 2s)^T$ where s is chosen so that $p_0^T g_1 = 0$, where

$$g_1 = \begin{pmatrix} 2 - 6s + 1 - 2s \\ 1 - 3s + 1 - 2s \end{pmatrix} = \begin{pmatrix} 3 - 8s \\ 2 - 5s \end{pmatrix}.$$

Hence

$$p_0^T g_1 = -3 \times (3 - 8s) - 2 \times (2 - 5s).$$

By solving $p_0^T g_1 = 0$ we get the perfect steplength and the new point as

$$s^* = \frac{13}{34} \text{ and } x_1 = \frac{1}{34}\begin{pmatrix} -5 \\ 8 \end{pmatrix} \quad \text{where} \quad g_1 = \frac{1}{34}\begin{pmatrix} -2 \\ 3 \end{pmatrix}.$$

We now use g_1 to find a search direction for the next iteration. First we get

$$\beta = \frac{g_1^T g_1}{g_0^T g_0} = \frac{13}{(34^2 \times 13)} = \frac{1}{34^2}$$

and then

$$p_1 = -g_1 + \beta p_0 = \frac{1}{34}\begin{pmatrix} 2 \\ -3 \end{pmatrix} + \frac{1}{34^2}\begin{pmatrix} -3 \\ -2 \end{pmatrix} = \frac{1}{34^2}\begin{pmatrix} 65 \\ -104 \end{pmatrix}.$$

The new solution estimate reached at the end of the second iteration will be

$$x_2 = x_1 + sp_1 = \left(-\frac{5}{34} + \frac{65s}{34^2},\ \frac{8}{34} - \frac{104s}{34^2}\right)^T$$

which gives

$$g_2 = \left(-\frac{2}{34} + \frac{26s}{34^2},\ \frac{3}{34} - \frac{39s}{34^2}\right)^T.$$

For a perfect line search the steplength s satisfies $p_1^T g_2 = 0$, i.e.,

$$-\frac{130}{34^3} + \frac{(65 \times 26s)}{34^4} - \frac{312}{34^3} + \frac{(104 \times 39s)}{34^4} = 0.$$

After simplification this leads to

$$s^* = 34 \times \frac{442}{5746} \approx 2.5562.$$

Thus, after two iterations, the conjugate gradient method has reached x_2 where

$$x_{2_1} = -\frac{5}{34} + \frac{(34 \times 442 \times 65)}{5746 \times 34^2} = 0 \text{ and } x_{2_2} = \frac{8}{34} - \frac{(34 \times 442 \times 104)}{5746 \times 34^2} = 0.$$

This point minimizes the function since $g(x_2) = 0$. Hence the example demonstrates the finite termination property of the conjugate gradient method applied to a quadratic function.

Exercises

1. Do two conjugate gradient iterations, starting from $x = (0,0)^T$, applied to

$$F(x) = 2x_1^2 + x_1x_2 + x_2^2 + x_1 - x_2.$$

What do you observe about the result?

2. When the conjugate gradient algorithm is applied to a quadratic function show that the steplength calculation (8.1.2) will ensure that $p_k^T g_{k+1} = 0$. Show also that this value of s can be found, without using A directly, from

$$s = -\frac{p_k^T g_k}{p_k^T (g^+ - g_k)}$$

where $g^+ = Ax^+ + b$ and $x^+ = x_k + p_k$.

3. Extend the proof of (8.1.5) to show that after k iterations

$$p_j^T g_k = 0 \quad \text{for } j = 0, 1, .., k-1.$$

How does this result lead to the finite termination property that the conjugate gradient method minimizes a quadratic function in at most n iterations?

4. Show that the search directions and gradients generated by the conjugate gradient algorithm satisfy *(i)* $g_{k+1}^T g_k = 0$; and *(ii)* $p_k^T g_k = -g_k^T g_k$. Hence show that the choice (8.1.1) for β ensures that p_{k+1} and p_k are conjugate w.r.t. A.

5. A quasi-Newton method with perfect line searches and using the *DFP* update is applied to a quadratic function F. Show that successive search directions are conjugate with respect to $\nabla^2 F$.

6. Show that the eigenvectors of a symmetric matrix A are also conjugate directions with respect to A.

2. Conjugate gradients and general functions

The conjugate gradient method can be used to minimize a positive definite quadratic function in at most n iterations from any starting point. We can also modify it as an algorithm for minimizing a general function $F(x)$. This algorithm proceeds in "cycles" of n iterations, with p_n, p_{2n}, etc being re-set as the steepest descent direction. This is because we cannot have more than n vectors which are mutually conjugate to a given matrix. It is therefore recommended that each cycle of n steps is regarded as making progress to the minimum of a local quadratic model of F. If this does not yield a suitable estimate of the true minimum then a fresh cycle must be started.

Conjugate gradient method for minimizing $F(x)$

Choose x_0 as an initial estimate of the solution
Calculate $g_0 = \nabla F(x_0)$. Set $p_0 = -g_0$
Repeat for $k = 0, 1, 2, ...$
find s by a perfect line search to minimize $F(x_k + sp_k)$
set $x_{k+1} = x_k + sp_k$, $g_k = \nabla F(x_k)$
if k is not a multiple of n then
find β from (8.1.1) and set $p_{k+1} = -g_{k+1} + \beta p_k$
else
set $p_{k+1} = -g_{k+1}$
until $||g_{k+1}||$ is sufficiently small.

The calculation of β from (8.1.1) is called the *Fletcher-Reeves* formula [32]. An alternative, due to *Polak & Ribiere* (*PR*) [33], is

$$\beta = \frac{g_{k+1}^T(g_{k+1} - g_k)}{g_k^T g_k}. \tag{8.2.1}$$

When F is quadratic (8.1.1) and (8.2.1) give the same β. When F is not quadratic, however, (8.1.1) and (8.2.1) will lead to different search directions. (Of course, when $F(x)$ is not quadratic, the search directions p_k, p_{k-1} are not truly conjugate since there is no constant Hessian $\nabla^2 F$ for them to be conjugate with respect to!)

Other formulae for obtaining conjugate search directions are given in [13].

Exercises

1. Show that, when applied to a general non-quadratic function, the conjugate gradient method *with perfect line searches* generates a descent direction on every iteration.

2. Show that the formulae (8.1.1) and (8.2.1) are equivalent when F is a quadratic function.

3. Convergence of conjugate gradient methods

We can establish convergence of the conjugate gradient method using Wolfe's Theorem. We can show that p_k is always a descent direction (Exercise 1 in the previous section) and the perfect line search ensures that (5.5.2), (5.5.3) hold.

In practice the conjugate gradient algorithm usually needs more iterations than a quasi-Newton method. Its ultimate rate of convergence is *n-step quadratic*, which means that

$$||x_k - x^*|| \leq C||x_{k-n} - x^*||^2$$

for some constant C and for k sufficiently large. This implies that convergence will usually be slower than for the Newton and quasi-Newton approaches.

Convergence of conjugate gradient methods can be accelerated by use of *preconditioning*. Prior to the solution of a system $Ax+b=0$, transformations can be applied to the matrix A to cause its eigenvalues to become closer together. This is to exploit a stronger finite termination property of the conjugate gradient method which states that the number of iterations required to solve $Ax+b=0$ will be bounded by the number of *distinct* eigenvalues of A. For more information on this and on the many other variants of the conjugate gradient approach see [34].

In spite of having slower convergence, the conjugate gradient method does have a potential advantage over Newton and quasi-Newton techniques. It does not use any matrices and therefore requires less computer memory when the number of variables, n, is large. Moreover the number of multiplications per iteration is $O(n)$, compared with $O(n^2)$ for the quasi-Newton method and $O(n^3)$ for the Newton approach.

4. Numerical results with conjugate gradients

The SAMPO implementations of the conjugate-gradient method use the Fletcher-Reeves recurrence (8.1.1) and are denoted by CGp and CGw, signifying, respectively, the use of perfect and weak linesearches. The theory behind the conjugate-gradient method makes it much more strongly dependent on the use of perfect searches than any of the other minimization techniques considered so far. Table 8.1 shows numbers of iterations and function calls needed by CGp and CGw to solve problems T1 –T4 by minimizing **Minrisk1m** or **Maxret1m** to standard accuracy (4.4.2). The figures in Table 8.1 were obtained with the SAMPO program sample3.

	T1 $\rho=100$	T2 $\rho=50$	T3 $\rho=2$	T4 $\rho=50$
CGp	4/13	4/10	30/284	66/439
CGw	34/71	8/18	44/163	84/252

Table 8.1. Performance of CGp and CGw on problems T1 – T4

Points to note from Table 8.1 are as follows.

• On the quadratic problems T1 and T2, CGp behaves like QNp and terminates in four iterations. This agrees with theoretical expectations for the four-variable quadratic problem **Minrisk1m**. If, however, we solve problems T1 and T2 using CGw then performance is not so good as that of QNw (see Table 7.1). The conjugate gradient approach is more sensitive to the accuracy of the linesearch.
• On the non-quadratic problems T3 and T4, both CGp and CGw are more expensive than the Newton and quasi-Newton approaches (Tables 6.1 and 7.1). CGw often does better than CGp in terms of function calls even though there is little theoretical justification for using weak linesearches. Conjugate gradient methods may have an advantage over Newton or quasi-Newton methods only if their reduced arithmetic cost per iteration can compensate for the extra iterations and function calls they require.

To illustrate the ultimate convergence of the conjugate gradient method we consider problem T4 and solve **Maxret1m** with $\rho = 50$. Table 8.2 shows the numbers of iterations and function calls needed by the conjugate gradient method and by the (superlinearly convergent) quasi-Newton approach to meet low- and high-accuracy convergence tests (4.4.3). The n-step quadratic convergence of the conjugate-gradient method means that, in practice, it can take significantly more iterations than a quasi-Newton technique to reduce the gradient norm by an extra order of magnitude.

	itns/fns (low)	itns/fns (high)
QNp	21/118	24/124
QNw	26/51	35/60
CGp	58/415	76/466
CGw	76/232	91/272

Table 8.2. Ultimate convergence of QN and CG on problem T4

Exercises (To be solved using `sample3` or other suitable software.)
1. Apply the conjugate-gradient method to the problem **Risk-Ret1** using the dataset `Real-20-5` with $\rho = 10$.

2. Apply the conjugate gradient method (with weak and perfect line searches) to **Maxret1m**, using data from Table 4.1. Use several values of V_a between 0.003 and 0.004 and try $\rho = 0.1, 1$ and 10. Do perfect line searches generally give more accurate solutions? Are perfect or weak line searches more computationally efficient?

3. How does the conjugate gradient method behave when applied to the maximum return problem in the form (7.4.1) using data from Table 1.2?

4. Produce an inplementation of the conjugate gradient method which uses the Polak-Ribiere formula (8.2.1) for β rather than the Fletcher-Reeves form. How does it perform on the problems in the first three questions?

5. The truncated Newton method

We can now describe an approach which combines the Newton and conjugate gradient methods. As explained in an earlier chapter, the essential feature of a Newton iteration is the calculation of a search direction, p, from the linear system, involving $G = \nabla^2 F$ and $g = \nabla F$,

$$Gp = -g. \quad (8.5.1)$$

However, the solution of (8.5.1) can be computationally expensive and the development of quasi-Newton methods was motivated by the wish to avoid forming and factorizing the exact Hessian.

It can also be argued that we could do less arithmetic and yet retain some benefits of the Newton method if we were to form G as the true Hessian matrix and then obtain p by only *approximately* solving (8.5.1). We would require this approximate solution to be significantly cheaper than, say, the Cholesky method for solving (8.5.1) exactly. Of course, we would also want the resulting vector p to be a useful search direction. One way of getting such an approximate solution is to apply the conjugate gradient method with a fairly large tolerance on the residuals $||Gp+g||$ so that the iteration terminates in appreciably fewer than n steps. The *Truncated Newton* approach introduced by Dembo, Eisenstat and Steihaug [35] makes use of this idea. We give below a version of this algorithm for minimizing a *convex* function $F(x)$.

Truncated Newton method for minimizing convex $F(x)$

Choose x_0 as an initial estimate of the solution
Choose C as a constant > 1
Repeat for $k = 0, 1, ...$
 Calculate $g_k = \nabla F(x_k)$ and $G_k = \nabla^2 F(x_k)$.
 Set $\nu_k = \min\{C||g_k||, k^{-1}\}$
 Apply conjugate gradient iterations to the system $G_k p = -g_k$
 and take p_k as the first solution estimate for which $||G_k p_k + g_k|| < \nu_k$.
 Find s so $(x_k + sp_k)$ satisfies Wolfe conditions 2 and 3 for some η_1, η_2
 Set $x_{k+1} = x_k + sp_k$
until $||g_k||$ is sufficiently small.

The algorithm differs from the standard Newton approach mainly in its use of a parameter ν_k which governs the accuracy with which the Newton equation is solved in order to obtain a search direction. The formula for ν_k means that it decreases as k increases and as the gradient g_k becomes smaller. Hence p_k tends to the Newton direction as the search gets nearer to an optimum and so the ultimate convergence can be expected to be fast. The potential benefit of the method lies in the fact that it costs less per iteration than the classical Newton technique while the search is still far from a minimum of F.

Of course, if the line search in the Truncated Newton method is to succeed, the search direction p_k must satisfy Wolfe condition 1 – i.e. it must be a descent direction. To ensure that this condition will hold, the (inner) conjugate gradient iterations must be as follows.

Conjugate gradient inner iteration for the Truncated Newton method.

Given G_k, g_k and η_k, set $u_0 = 0$, $v_0 = -g_k$ and $r_0 = g_k$
Repeat for $j = 1,2,..,n$
set $u_j = u_{j-1} + sv_{j-1}$, $r_j = G_k u_j + g_k$ where

$$s = \frac{-v_{j-1}^T r_{j-1}}{v_{j-1}^T G_k v_{j-1}}$$

if $||r_j|| < \nu_k$ then set $p_k = u_j$ and exit
if $u_j^T g_k \geq 0$ then set $p_k = u_{j-1}$ and exit
set $v_j = -r_j + \beta v_{j-1}$ where

$$\beta = \frac{r_j^T r_j}{r_{j-1}^T r_{j-1}}$$

Because the above algorithm will ensure that u_1 is parallel to $-g_k$ it follows that – at worst – p_k will be the negative gradient and so the inner iteration must return a descent direction.

If the Truncated Newton method is applied to a non-convex function $F(x)$ then the matrix G_k may not be positive definite on some iterations. To cater for this situation we need to provide another exit from the inner conjugate gradient process. If, during the calculation of the step-length, s, the denominator is found to be negative or zero then this indicates that the Hessian matrix is locally non-positive-definite and hence that the conjugate gradient method is not suitable for an accurate solution of the Newton equation. Hence we require an extra test at the start of the inner iteration loop, namely:

if $v_{j-1}^T G_k v_{j-1} \leq 0$ then
if $j = 1$ set $p_k = -g_k$ and exit
else set $p_k = u_{j-1}$ and exit

Saddle point

You thought you'd arrived
till the valley arched its back
like a startled cat.

Chapter 9

OPTIMAL PORTFOLIOS WITH RESTRICTIONS

1. Introduction

In the examples presented so far, optimal portfolios have been allowed to include any amount of short-selling. For instance, the solution to problem T3 has $y_3 \approx -0.3$. We now consider how to find minimum-risk and maximum-return portfolios when short-selling is not permitted. We can write these problems as

Minrisk2

$$\text{Minimize} \quad V = y^T Q y$$

$$\text{subject to} \quad \sum_{i=1}^{n} \bar{r}_i y_i = R_p \quad \text{and} \quad \sum_{i=1}^{n} y_i = 1 \quad \text{and} \quad y_i \geq 0, \ i = 1,..,n.$$

Maxret2

$$\text{Minimize} \quad R = -\bar{r}^T y$$

$$\text{subject to} \quad V = y^T Q y = V_a \quad \text{and} \quad \sum_{i=1}^{n} y_i = 1 \quad \text{and} \quad y_i \geq 0, \ i = 1,..,n.$$

Solution methods which handle these constrained optimization problems directly will be considered in later chapters. However, in section 2 below we shall show how to use a transformation of variables to exclude the possibility of short-selling. To prepare the way for this, we first introduce another way of dealing with **Minrisk1** and **Maxret1**. Consider the *unconstrained* problem

Minrisk1u

$$\text{Minimize} \quad F(y) = \frac{\rho}{R_p^2}(R - R_p)^2 + \rho(S-1)^2 + V. \qquad (9.1.1)$$

In (9.1.1), $V = y^T Qy$, $R = \bar{r}^T y$ and $S = e^T y$ and so the minimum of $F(y)$ can be expected to occur at a point which *approximates* the solution to **Minrisk1**. The value of V will be small and, when ρ is large enough, the first two terms in $F(y)$ will be close to zero and so equations (1.2.3) and (1.2.2) will be close to being satisfied. Solving (9.1.1) means we are working directly with the invested fractions y_i and we avoid the use of the transformations involving (3.1.5).

In the expected return term in **Minrisk1u** (and also in **Minrisk1m**) we could omit the denominator R_p^2. The objective function would then be

$$\hat{F}(y) = \rho(R - R_p)^2 + \rho(S-1)^2 + V.$$

Now suppose that the invested fractions y_i are such that $|R - R_p| \approx 0.1R_p$ and $|S-1| \approx 0.1$ so that the *percentage* errors in both constraints are roughly equal (at about 10%). If R_p is quite small – say 0.1% – then the first term in $\hat{F}(y)$ will be about one hundred times smaller than the second. This suggests that the condition $S = 1$ may have a disproportionate effect on the position of the minimum of $\hat{F}(y)$. With the scaled form for the expected-return term in (9.1.1), however, the return and normalisation contributions are better balanced.

In order to use a gradient method to solve (9.1.1) we need the derivatives of F with respect to the variables $y_1, ..., y_n$. Setting $\bar{\rho} = \rho/R_p^2$, we can write

$$\frac{\partial F}{\partial y_i} = 2\bar{\rho}(R - R_p)\frac{\partial R}{\partial y_i} + 2\rho(S-1)\frac{\partial S}{\partial y_i} + \frac{\partial V}{\partial y_i}.$$

From the definitions of R, S and V we get

$$\frac{\partial R}{\partial y_i} = \bar{r}_i, \quad \frac{\partial S}{\partial y_i} = 1 \quad \text{and} \quad \frac{\partial V}{\partial y_i} = 2\sum_{j=1}^{n} \sigma_{ij} y_j.$$

Thus the gradient vector F_y of the function in (9.1.1) is given by

$$F_y = 2\bar{\rho}(R - R_p)\bar{r} + 2\rho(S-1)e + 2Qy, \tag{9.1.2}$$

where e denotes the column vector $(1, 1, 1..., 1)^T$.

We can also pose **Maxret1** in a similar form, i.e.,

Maxret1u

$$\text{Minimize} \quad F(y) = \rho(S-1)^2 + \frac{\rho}{V_a^2}(V - V_a)^2 - R \tag{9.1.3}$$

where $V = y^T Qy$, $R = \bar{r}^T y$ and $S = e^T y$. Reasons for scaling the risk term in (9.1.3) are similar to those relating to the scaling of the expected return term in

(9.1.1). It is left to the reader to show that the gradient of $F(y)$ in (9.1.3) is

$$\frac{\partial F}{\partial y_i} = 2\rho(S-1) + 2\frac{\rho}{V_a^2}(V - V_a)\sum_{j=1}^{n}\sigma_{ij}y_j - \bar{r}_i. \qquad (9.1.4)$$

Exercises

1. Obtain an expression for the Hessian matrix F_{yy} of the function in (9.1.1).

2. Re-formulate problem **Minrisk0** in a similar way to (9.1.1).

3. Apply the ideas of this section to the problem **Risk-Ret1** and obtain expressions for the gradient and Hessian of the resulting objective function.

4. Verify that the expression (9.1.4) gives the correct gradient of (9.1.3) and obtain an expression for the second derivative matrix.

Comparing versions of Minrisk1 and Maxret1

The SAMPO software includes a program `sample4` which solves portfolio optimization problems in the form **Minrisk1u** and **Maxret1u**. Table 9.1 shows the expected return and risk at the solutions of problems T2 and T4 obtained from the 'm' and 'u' versions of **Minrisk1** and **Maxret1**. For smaller values of the parameter ρ the two formulations yield somewhat different results; but as ρ increases both versions produce R and V values which become closer to the correct solutions given in section 4 of chapter 4.

	T2				T4			
	Minrisk1m		**Minrisk1u**		**Maxret1m**		**Maxret1u**	
ρ	R	V	R	V	R	V	R	V
1	0.243	0.141	0.237	0.123	0.287	0.1638	0.288	0.1622
10	0.249	0.1423	0.249	0.1402	0.27	0.1519	0.272	0.1517
100	0.25	0.1425	0.25	0.1423	0.267	0.1502	0.268	0.1502

Table 9.1. Solutions from 'm' and 'u' forms of **Minrisk1** and **Maxret1**

Exercises (To be solved using `sample4` or other suitable software.)

1. For problem T1, what is the smallest value of ρ that causes the solutions of **Minrisk1m** and **Minrisk1u** to agree to two significant figures? Is the same value of ρ sufficient to ensure that **Maxret1m** and **Maxret1u** give comparable solutions for problem T3?

2. Find how many iterations and function calls are needed by the Newton, quasi-Newton and conjugate gradient methods to solve problems T1 and T2

via **Minrisk1u** and problems T3 and T4 via **Maxret1u** (using $\rho = 100$ in all cases). How do these figures compare with those for **Minrisk1m** or **Maxret1m** with $\rho = 100$? Would you expect the 'u' problems to be more difficult than the 'm' problems?

2. Transformations to exclude short-selling

The problems **Minrisk1u** and **Maxret1u** do not, in themselves, offer any very significant advantages over **Minrisk1m** and **Maxret1m**. They are important because they prepare the way for other forms of minimum-risk and maximum-return problem which exclude the possibility of short-selling.

We begin with the function (9.1.1) and introduce new variables $x_1,..,x_n$ such that $y_i = x_i^2$. *Note that this transformation ensures that the invested fractions y_i are non-negative.* Expressing F as a function of $x_1,..,x_n$ is straightforward and we obtain the problem

Minrisk2u

$$\text{Minimize } \frac{\rho}{R_p^2}(\sum_{i=1}^{n} \bar{r}_i x_i^2 - R_p)^2 + \rho(\sum_{i=1}^{n} x_i^2 - 1)^2 + \sum_{i=1}^{n} x_i^2 (\sum_{j=1}^{n} \sigma_{ij} x_j^2). \quad (9.2.1)$$

We denote the function in (9.2.1) by $F(x)$. To solve **Minrisk2u** by a gradient method, we need to calculate the first derivatives of F with respect to $x_1,..,x_n$. Considering F as a function of y and y as a function of x, we can write

$$\frac{\partial F}{\partial x_i} = \sum_{j=1}^{n} \frac{\partial F}{\partial y_j} \frac{\partial y_j}{\partial x_i}.$$

However the relationship between y and x means that

$$\frac{\partial y_j}{\partial x_i} = \begin{cases} 2x_i & \text{if } i = j \\ 0 & \text{if } i \neq j \end{cases}$$

We already know from section 1 of this chapter that, with $\bar{\rho} = \rho/R_p^2$,

$$\frac{\partial F}{\partial y_j} = 2\bar{\rho}(R - R_p)\bar{r}_j + 2\rho(S-1) + 2\sum_{k=1}^{n} \sigma_{jk} y_k$$

and therefore

$$\frac{\partial F}{\partial x_i} = 4x_i[\bar{\rho}(R - R_p)\bar{r}_i + \rho(S-1) + 2\sum_{k=1}^{n} \sigma_{ik} x_k^2] \quad (9.2.2)$$

where

$$R = \sum_{k=1}^{n} \bar{r}_k x_k^2 \text{ and } S = \sum_{k=1}^{n} x_k^2.$$

Clearly, (9.1.3) can also be re-written via transformation $y_i = x_i^2$, $(i = 1,...,n)$ to give a problem involving $x_1,...,x_n$ as optimization variables, namely

Maxret2u

$$\text{Minimize } \frac{\rho}{V_a^2}(\sum_{i=1}^{n} x_i^2 - 1)^2 + \rho(\sum_{i=1}^{n} x_i^2(\sum_{j=1}^{n} \sigma_{ij}x_j^2) - V_a)^2 - \sum_{i=1}^{n} \bar{r}^T x_i^2. \qquad (9.2.3)$$

We leave it to the reader to show that if $F(x)$ is the function in (9.2.3) then

$$\frac{\partial F}{\partial x_i} = 4x_i[-\bar{r}_i + \rho(S-1) + 2\frac{\rho}{V_a^2}(V - V_a)\sum_{j=1}^{n} \sigma_{ij}x_j^2]. \qquad (9.2.4)$$

Exercises

1. Obtain expressions for the second derivatives of (9.2.1).

2. Verify that the first derivative expression (9.2.4) is correct. Hence obtain expressions for the second derivatives of (9.2.3).

3. Using the ideas from this section, modify the problem **Risk-Ret1** to obtain a variant which excludes short-selling.

3. Results from Minrisk2u and Maxret2u

The `SAMPO` program `sample4` can solve problems **Minrisk2u** and **Maxret2u** by a number of different optimization methods. It employs the same standard starting guess and convergence tests as `sample3`. In this section we shall quote results obtained with `sample4` for the following test problems.

Problem T5 involves data from Table 1.1 – for which $\bar{r}$ and Q are given by (4.4.10) and (4.4.11) – and seeks a minimum-risk portfolio for target return 1.25%. The solution is given by

$$y_1 \approx 0.76, \; y_2 \approx 0.41, \; y_3 \approx -0.18 \qquad (9.3.1)$$

with minimum risk $V \approx 1.8 \times 10^{-3}$.

Problem T6 is a maximum-return problem using the data in Table 4.1, for which $\bar{r}$ and Q are given by (4.4.4), (4.4.5). The acceptable risk is $V_a = 0.0075$. The invested fractions which solve this problem are

$$y_1 \approx 0.31, \; y_2 \approx 0.39, \; y_3 \approx 0.02, \; y_4 \approx 0.43, \; y_5 \approx -0.15 \qquad (9.3.2)$$

giving an expected return $R \approx 1.19\%$.

Problems T1a–T6a are the same as problems T1–T6 but have the additional restriction that no short-selling is permitted.

Solving Minrisk2u

Table 9.2 compares the solution to problem T5 from **Minrisk1u** with that for problem T5a obtained from **Minrisk2u**. Clearly the transformation $y_i = x_i^2$ in **Minrisk2u** has worked as intended and excluded short-selling. However the risk associated with the optimal portfolio for problem T5a is appreciably greater than that for T5.

Problem	ρ	y	R	V
T5 (**Minrisk1u**)	100	0.76, 0.41, -0.17	1.25%	0.0018
T5a (**Minrisk2u**)	100	0.78, 0.22, 0	1.25%	0.0038

Table 9.2. Solutions for problems T5 and T5a

Minrisk2u is a more difficult optimization problem than either **Minrisk1m** or **Minrisk1u** because it involves a non-quadratic function. Hence the Newton method cannot be expected to solve it in one iteration and `QNp` is not guaranteed to converge within n iterations. Table 9.3 shows how many iterations and function calls are needed by the Newton, quasi-Newton and conjugate-gradient methods to solve problems T1a and T2a via **Minrisk2u**.

	T1a $\rho = 100$	T2a $\rho = 50$
`NMp`	15/95	11/38
`NMw`	73/101	19/23
`QNp`	118/675	23/106
`QNw`	247/387	45/72
`CGp`	350/1493	83/303
`CGw`	461/1204	145/379

Table 9.3. Comparing `NM`, `QN` and `CG` on problems T1a and T2a

`NMw` gives the most rapid convergence while `QNw` comes a not particularly close second. The conjugate-gradient method does not seem at all competitive on either problem.

Exercises (To be solved using `sample4` or other suitable software.)

1. By solving **Minrisk1m** and **Minrisk2u**, determine two efficient frontiers for the assets in Table 4.1 – one with short-selling and one without.

2. For the assets in Table 1.1, what is the largest value of R_p which can be obtained without short-selling?

3. Use **Minrisk2u** to solve mimimum-risk problems with the data for problem T5a but with $R_p = 1.3\%$ and $R_p = 1.4\%$. What do you notice about the solutions? *This example should serve as a warning that it is quite easy to impose restrictions in an optimization problem in such a way that there is no feasible solution.*

4. Compare the performance of the Newton, quasi-Newton and conjugate gradient methods to solve problem T5a via **Minrisk2u**.

Solving Maxret2u

Table 9.4 shows the solutions obtained for problems T6 (with **Maxret1u**) and T6a (using **Maxret2u** to exclude short-selling).

Problem	ρ	y	R	V
T6 (**Maxret1u**)	100	0.31, 0.39, 0.02, 0.43, -0.15	1.19%	0.0075
T6a (**Maxret2u**)	100	0.37, 0.63, 0, 0, 0	1.16%	0.0075

Table 9.4. Solutions for problems T6 and T6a

Once again, the $y_i = x_i^2$ transformation has worked as intended. Forcing the invested fractions to be non-negative has eliminated three assets from the portfolio and reduced the maximum return obtainable for the specified risk.

We would expect **Maxret2u** to be harder to solve than **Maxret1m** or **Maxret1u** because the $y_i = x_i^2$ transformation increases the nonlinearity of an already non-quadratic objective function. Table 9.5 shows numbers of iterations and function calls needed by the Newton, quasi-Newton and conjugate-gradient methods to solve problems T3a and T4a via **Maxret2u**.

	T3a $\rho = 2$	T4a $\rho = 50$
`NMp`	16/79	15/79
`NMw`	15/29	20/23
`QNp`	21/124	47/255
`QNw`	28/44	72/116
`CGp`	64/322	75/392
`CGw`	86/220	124/360

Table 9.5. Comparing `NM`, `QN` and `CG` on problems T3a and T4a

Comparison with Tables 6.1 and 7.1 confirms that solving **Maxret2u** can be more expensive than solving **Maxret1m**. As with the results in Table 9.3, we find that NMw is usually the most efficient method. In terms of function evaluations, QNw beats NMp into second place on problem T3a. Once again, the conjugate-gradient approach comes quite a poor third.

Exercises (To be solved using sample4 or other suitable software.)
1. For the dataset Real-20-5 calculate and plot values of maximum return against acceptable risk by solving **Maxret2u** for $0.14 \leq V_a \leq 0.4$. How do the results compare with those given by solving **Maxret1m** and **Maxret1u**?

2. Explain what happens if you solve **Maxret2u** starting from $x_1, .., x_n = 0$.

3. Compare numbers of iterations and function calls used by the Newton, quasi-Newton and conjugate gradient methods to solve T6a via **Maxret2u**.

4. Upper and lower limits on invested fractions

As well as excluding short-selling, we might also want to put some upper limit on the y_i. For instance we could choose not to commit more than 25% of our investment to any single asset. To help us to build such a condition into a minimum risk problem we let y_L and y_U be, respectively, the lower and upper bounds on the invested fractions and then introduce the transformation

$$y_i = y_L + (y_U - y_L) sin^2 x_i. \tag{9.4.1}$$

Clearly this will give $y_L \leq y_i \leq y_U$ for *any* value of x_i. Hence, if we subsititute (9.4.1) into (9.1.1) we obtain a version of the minimum risk problem which imposes bounds on the invested fractions. It is left to the reader to complete the formulation of this problem and to obtain derivatives of the objective function.

Another way of extending **Minrisk1u** to handle bounds on the investment fractions involves adding extra penalty terms into (9.1.1). In the case of upper bounds, for instance, we can introduce a function

$$\nu(y_i,\, y_U) = \begin{cases} 0 & \text{if } y_i \leq y_U \\ (y_i - y_U)^2 & \text{if } y_i > y_U \end{cases} \tag{9.4.2}$$

We can now extend (9.1.1) as

$$\text{Minimize} \quad F(y) = \frac{\rho}{R_p^2}(R - R_p)^2 + \rho(S-1)^2 + V + \rho \sum_{i=1}^{n} \nu(y_i, y_U). \tag{9.4.3}$$

The solution of (9.4.3) can be expected to occur at a point which gives a small value of risk while producing an expected return close to R_p *and* causing the y_i to be below – or at least not too much above – the limit y_U.

The reader should easily be able to obtain expressions for the first derivatives of the function $F(y)$ given by (9.4.3).

Exercises

1. Write down the function of x which results from using the substition (9.4.1) in (9.1.1). Hence obtain the first and second partial derivatives of this function with respect to x_i, $i = 1, ..., n$.

2. Obtain a variant of problem **Maxret1u** which uses (9.4.1) to impose bounds on the investment fractions. Also obtain expressions for the elements of the gradient and Hessian for this problem.

3. Determine the gradient and Hessian matrix of the function (9.4.3). Do the same for an extension of **Maxret1u** which applies both lower and upper bounds on the variables y_i.

4. By modifying the program `sample4` (or using other suitable software) implement the transformation (9.4.1) and hence solve problems T2 and T4 subject to limits $0.01 \leq y_i \leq 0.3$ for $i = 1, .., 5$.

Trust region

How far dare I go?
I've a hunch how you'll react
but I might be wrong.

I want to tell you
where my fancy's heading but
I must be gentle

for to alarm you
would risk ending the affair.
before it's started.

Chapter 10

LARGER-SCALE PORTFOLIOS

1. Introduction

The examples in this chapter are more realistic than most of those considered previously. We have already mentioned that the SAMPO includes data on asset performance from the London Stock Market in 2002-03. We can also make pseudo-random perturbations to this data in order to generate synthetic histories of returns. In this way we can set up large-scale portfolio optimization problems involving many assets and long performance histories. We use Synth-m-n to denote a synthetic dataset involving n assets and an m-day history. In the next section we consider a set of such artificial problems in order to compare the performance of optimization methods as the number of variables increases. In the second part of the chapter we use the genuine market data to illustrate some practical points about the behaviour of optimized portfolios.

We shall use the following families of test problems.
Problem T7(n) uses dataset Synth-400-n and involves finding the minimum-risk portfolio for $R_p = 2R_{min}$, where R_{min} is the expected return at the solution of problem **Minrisk0**.
Problem T8(n) also uses Synth-400-n and involves finding the maximum-return portfolio with acceptable risk $V_a = 2V_{min}$, where V_{min} is the portfolio risk at the solution of **Minrisk0**.
Problem T9(m) uses dataset Real-m-5 and seeks the minimum-risk portfolio to produce an expected return $R_p = 0.1\%$.
Problem T10(m) uses dataset Real-m-5 and seeks the portfolio giving maximum return for a risk $V_a = 1.1V_{min}$, where V_{min} is the least possible value of risk, obtained by solving **Minrisk0**.

2. Portfolios with increasing numbers of assets

Minimum-risk problems

Table 10.1 shows numbers of iterations and function calls needed to solve problem T7 for various n. These results are obtained by applying `sample3` to problem **Minrisk1m** with $\rho = 10$. The standard convergence test (4.4.2) is used *except* for the last row of the table which shows the cost of CGp in reaching a low-accuracy solution (4.4.3).

	$n = 50$	$n = 100$	$n = 150$	$n = 200$
NMw	1/2	1/2	1/2	1/2
QNw	27/42	36/42	42/62	53/73
QNp	25/54	35/75	40/89	51/111
CGp	51/124	129/365	111/300	170/470
CGp(low)	38/92	58/152	64/175	109/296

Table 10.1. Performance of NM, QN and CG on problem T7(n)

We can make the following comments on the figures in Table 10.1.

• The fact that NMw converges in just one iteration in every case is not surprising since **Minrisk1m** is a quadratic problem. What is more interesting is that both QNp and CGp converge in appreciably fewer than n iterations. A reason for this faster-than-expected convergence appears to be that the objective function in **Minrisk1m** typically has a rather "flat" optimum – i.e., there is quite a large region around the minimum within which the function value does not change very rapidly. Once the iterative search has located a point within this region then the convergence test can be satisfied quite easily.

• The last two rows of Table 10.1 show that ultimate convergence of CGp can be quite slow, even on a quadratic problem like **Minrisk1m**. At least $\frac{n}{4}$ iterations are needed to reduce the gradient norm from $O(10^{-4})$ to $O(10^{-5})$.

• In terms of iteration count, the Newton method is by far the most efficient method, while CGp is some 2-3 times worse than QNp for the larger values of n.

Exercises (To be solved with `sample3` or `sample4` or other suitable software.)

1. Use the results in Table 10.1 to predict how many iterations QNw will need to solve problem T7(n) with $n = 250, 300$ and 350. Check your predictions by actual solution of the problems. What do you observe about the minimum risk as n increases?

2. For $50 \leq n \leq 200$, compare numbers of iterations and function calls needed by various optimization methods to solve Problem T7(n) via **Minrisk2u**.

Maximum return problems

Table 10.2 shows numbers of iterations and function calls needed to solve Problem T8 for various values of n by running program `sample3` on problem **Maxret1m** with $\rho = 10$. With the exception of the last row, the figures relate to the standard convergence test (4.4.2).

	$n = 50$	$n = 100$	$n = 150$	$n = 200$
`NMw`	12/15	12/14	12/14	12/14
`QNw`	48/81	58/81	70/127	93/169
`QNp`	38/214	46/216	60/252	70/309
`CGp`	296/1440	395/2006	599/3082	811/3874
`CGp`(low)	227/1241	329/1686	455/2617	569/3109

Table 10.2. Performance of `NM`, `QN` and `CG` on problem T8(n)

We can make the following observations about Table 10.2.

- **Maxret1m** is not a quadratic problem and so we no longer expect the Newton method to converge in one iteration. Moreoever it would not be surprising if the quasi-Newton and conjugate gradient methods were to need more than n iterations. In fact, performance of `NMw` remains fairly consistent for all values of n. The number of iterations needed by `QNp` and `QNw` does increase steadily with n, but always lies between n and $\frac{n}{2}$. `CGp`, however, is unable to converge to either low or standard accuracy in less than about $3n$ iterations.
- The behaviour of the conjugate gradient method (as in Table 10.1) shows that it is quite sensitive to rounding errors. The conjugacy and orthogonality properties on which the algorithm depends will not be achieved precisely in finite precision arithmetic. This is particularly noticeable when there are large numbers of variables, because this tends to increase the amount of round-off in key calculations like the computation of scalar products. Hence the convergence and termination properties predicted by theory are not necessarily observed in practice.

Overhead costs and run-times

As n increases, it is relevant to compare methods not only on the basis of numbers of iterations and function calls but also in terms of the *time* they need to find a solution. This will depend on the *overhead costs* incurred on each iteration – i.e. the work done in computing a search direction, updating second derivative information etc.

For large n, the main cost of an iteration of the quasi-Newton method in SAMPO is about $3.5n^2$ multiplications to calculate a search direction and update the inverse Hessian estimate. The figures for problem T8 in Table 10.2 show that each iteration of QNw uses about 1.5 function calls; and the dominant cost of a function evaluation for both **Minrisk1m** and **Maxret1m** is the n^2 multiplications needed to calculate the risk, $V = y^T Qy$. Therefore the total cost per iteration of QNw can be estimated as about $5n^2$ multiplications. Similarly for QNp, which makes about 4.5 function calls per iteration, the total cost per iteration is given by $8n^2$ multiplications.

The overheads of CGp are linear in n because it does no matrix calculations and so the dominant cost is due to the function evaluations. Hence, from Table 10.2, we estimate that CGp uses about $4.7n^2$ multiplications per iteration.

The SAMPO implementation of NMw uses about $\frac{1}{3}n^3 + \frac{1}{2}n^2$ multiplications to form the Hessian matrix and solve the Newton equation $Gp = -g$. For problem T8 it makes only about 1.1 function calls per iteration and so the total cost per iteration can be estimated by $(\frac{1}{3}n + 1.6)n^2$.

From these approximate counts of arithmetic operations we can deduce relative run-times of the various methods on problem T8 when n is large. For instance

$$\frac{\text{time for QNw}}{\text{time for NMw}} \approx \frac{5I_{qnw}}{(\frac{1}{3}n + 1.6)I_{nmw}}$$

where I_{qnw}, I_{nmw} denote the numbers of iterations needed for convergence by the quasi-Newton and Newton methods, respectively. For $n = 200$ we have $I_{qnw} = 93$ and $I_{nmw} = 12$ and so

$$\frac{\text{time for QNw}}{\text{time for NMw}} \approx \frac{5 \times 93}{68 \times 12} \approx 0.56.$$

This implies that QNw will solve the problem faster than NMw even though it uses appreciably more iterations and function calls. Similar analysis also shows that, for problem T8 with $n = 200$,

$$\frac{\text{time for QNp}}{\text{time for NMw}} \approx 0.68 \quad \text{and} \quad \frac{\text{time for CGp}}{\text{time for NMw}} \approx 4.6$$

suggesting that QNp will take less time than NMw but that CGp will be significantly slower. The conjugate gradient method fails to gain a time advantage over the other techniques because the large number function calls it needs are enough to outweigh its low overhead cost per iteration.

The predictions given in this section for relative run-times for the methods agree quite well with experimental measurements.

Exercises (To be solved with `sample3` or `sample4` or other suitable software.)
1. Run `NMp` on problem T8(n) and add the results to the comparisons in this section.

2. Use the estimates in this section to predict the ratio of `CGp` iterations to `NMw` iterations which will cause `CGp` to solve problem T8(500) faster than `NMw`. Obtain similar predictions for `QNp` and `QNw`.

3. For $100 \leq n \leq 200$, compare the performance of various optimization methods to solve problem T8(n) via **Maxret2u**. *If possible*, also compare execution times of the methods.

3. Time-variation of optimal portfolios

We now turn to problems based on real-life data. For a particular set of assets, we investigate how the optimal values of invested fractions are affected by m, the number of days in the return-history which generates $\bar{r}$ and Q. For the problems considered below we have taken m in the range $50 \leq m \leq 80$. All histories start from the same day 1 and so if we have found an optimal portfolio with $m = 50$ then the solution of a similar problem with $m = 55$ can be thought of as a *re-balancing* of the original portfolio, taking into account of asset performance over five more trading days.

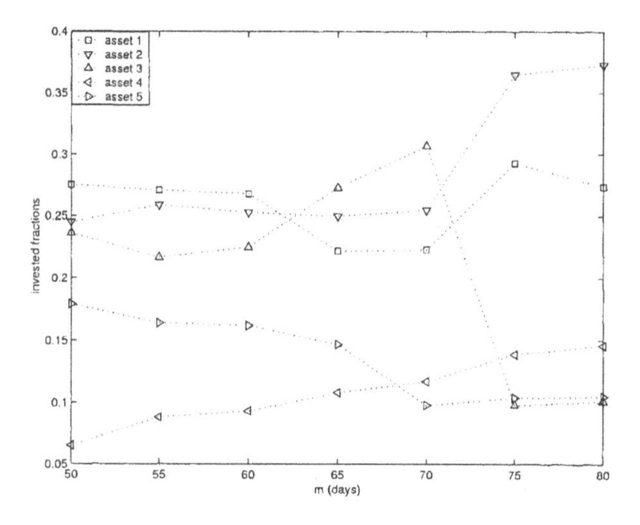

Figure 10.1. Solutions to problem T9 varying with time

Figure 10.1 shows how solutions of problem T9 can vary as the length, m, of asset performance history increases. (These solutions are obtained by forming and solving **Minrisk1m** by Newton's method.) For the first ten days or so, the optimal portfolio remains fairly constant; but then the balance between the assets begins to change. After twenty days the amount invested in asset two is much less than before and the fraction allocated to asset three is correspondingly increased.

Table 10.3 shows both the minimum-risk and the ratio V/V_{min} for the optimal portfolios from Figure 10.1. We can see that events occurring after day 70 must have significantly changed the expected returns and/or the variance-covariance matrix because the risk level increases appreciably.

	$m = 50$	$m = 55$	$m = 60$	$m = 65$	$m = 70$	$m = 75$	$m = 80$
V	0.524	0.535	0.499	0.502	0.509	0.662	0.619
V/V_{min}	1.012	1.012	1.014	1.063	1.082	1.043	1.029

Table 10.3. Risk levels at solutions to problem T9

Figure 10.2 shows solutions to problem T10(m) for $50 \leq m \leq 80$. These results were obtained by applying Newton's method to **Maxret1m**. Once again the invested fractions remain fairly constant for about ten days but then the optimal portfolios show a steady decrease in holdings of asset five and a very sharp reduction of investment in asset one. This is mostly compensated for by an increase in invested fraction y_2.

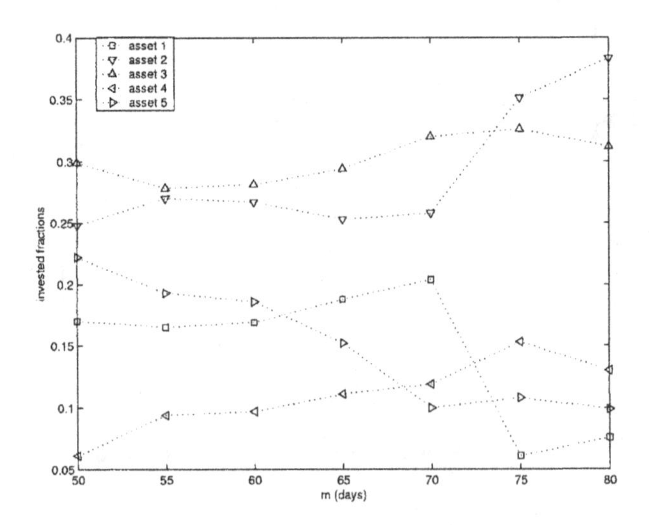

Figure 10.2. Solutions to problem T10 varying with time

Table 10.4 shows the expected returns for the portfolios illustrated in Figure 10.2. There is a significant change in optimal portfolio performance as asset history beyond day 70 is taken into account.

	$m=50$	$m=55$	$m=60$	$m=65$	$m=70$	$m=75$	$m=80$
$R\%$	0.185	0.188	0.179	0.125	0.114	0.245	0.222

Table 10.4. Portfolio returns at solutions to problem T10

The results quoted in this section indicate that it may be beneficial to re-balance a portfolio from time to time. In practice, however, *transaction costs* are incurred whenever assets are bought and sold and this means that it might not be efficient to make small adjustments to a portfolio such as those in Figures 10.1 and 10.2 when $50 \leq m \leq 60$. On the other hand, substantial changes in optimal invested fractions like those in Figures 10.1 and 10.2 for $m > 70$ probably would be worth implementing.

Exercises (To be solved with `sample3` or other suitable software.)
1. Investigate how solutions to problems T9 and T10 vary with m for the range $100 \leq m \leq 130$.

2. Gather some market data of your own for a group of 5-10 assets with a history of, say, 100 days and perform a similar investigation of the changes due to re-balancing portfolios.

4. Performance of optimized portfolios

We consider first the solution of problem T9 when $m = 50$. Using the historical database in `SAMPO`, we can calculate the return that *would have been* obtained from this optimal portfolio on subsequent days. From this we can obtain the mean portfolio return after m days where $m = 51, 52.., 100$. (This calculation is implemented as a post-optimization facility in `SAMPO`.) The history of mean returns is plotted in Figure 10.3 which shows that the reurn from the portfolio fluctuates about the target figure 0.1%. In this particular case, the portfolio does better than expectations over much of the period considered. It should be noted, however, that even optimal portfolios can lose money both in the long- and short-term (e.g. in Figure 10.3 for m between about 62 and 72).

It is important to understand that, irrespective of the history of past performance, if the assets in a portfolio are performing badly *now* then solving a minimum-risk or a maximum-return problem can still result in a loss-making situation over the short term, at least. Exercise 3 below illustrates this.

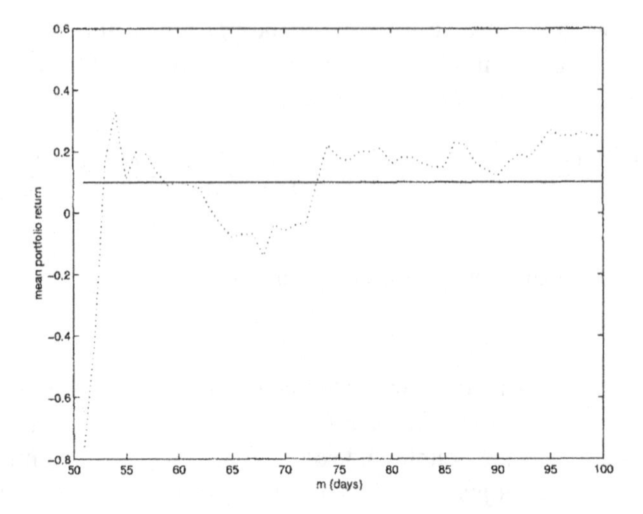

Figure 10.3. Mean returns from optimum portfolio for problem T9

Figure 10.4 shows a similar history of the performance of the portfolio which solves the maximum-return problem T10 with $m = 50$. The expected portfolio return in this case is $R \approx 0.185\%$.

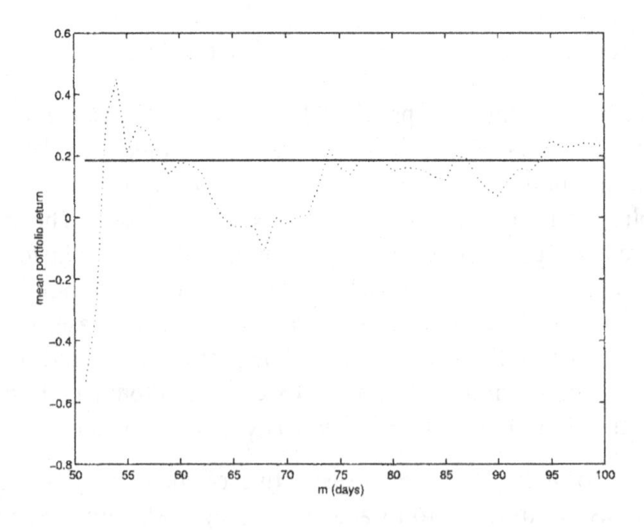

Figure 10.4. Mean returns from optimum portfolio for problem T10

Exercises (To be solved with `sample3` or `sample4` or other suitable software.)
1. How would the graphs of mean portfolio return in Figures 10.3 and 10.4 change if the portfolio were rebalanced every 10 days?

2. Investigate the behaviour of the optimal portfolios for problems T9 and T10 with $m = 50$ over a longer period of 50 days. Does periodic re-balancing improve the growth in portfolio worth?

3. Produce figures similar to 10.3 and 10.4 based on solutions to problems T9 and T10 with $m = 100$.

4. Perform similar experiments with real-life data to investigate the performance of optimal portfolios based on solving **Minrisk2u** and **Maxret2u**.

Time series (1)

A row of shark's teeth
jagged as a bandsaw to
cut off my prospects.

Chapter 11

DATA-FITTING & THE GAUSS-NEWTON METHOD

1. Data fitting problems

In *data fitting problems* we try to find a mathematical expression or *model* which matches a set of observations or measurements. For example, suppose we monitor the price of an asset at weekly intervals over a year, so that at week number i we have a value P_i (i = 1,...,52). Suppose also that, beneath day-to-day fluctuations, we believe the price shows an underlying periodic behaviour and so there is an approximate relationship

$$P(t) = x_1 + x_2 \sin(x_3 t + x_4) \tag{11.1.1}$$

where t is time (measured in weeks) and $x_1,..,x_4$ are unknown coefficients. We want to determine these coefficients so as to get the best match between the model (11.1.1) and the observed data. We would like $x_1,..,x_4$ to satisfy

$$P_i = x_1 + x_2 \sin(x_3 i + x_4) \quad \text{for} \quad i = 1,..,52.$$

However, we shall probably not be able to find four unknowns to satisfy all fifty-two equations. Therefore we look instead for the values of the x_k which will make the residuals

$$f_i(x) = P_i - (x_1 + x_2 \sin(x_3 i + x_4))$$

as small as possible. One way of doing this is to choose $x_1,..,x_4$ to minimize the two-norm of the residual vector $f = (f_1(x),..,f_{52}(x))^T$,

$$||f||_2 = \sqrt{(\sum_{i=1}^{52} f_i(x)^2)}.$$

For simplicity we can omit the square root and then we have the problem

$$\text{Minimize} \quad F(x) = \sum_{i=1}^{52} f_i(x)^2.$$

This approach is often called *least-squares* data fitting. We could of course try to match a set of data by using other norms, e.g. the L_1 norm, as in

$$\text{Minimize} \quad \sum_{i=1}^{m} |f_i(x)|$$

or the L_∞ norm as in

$$\text{Minimize} \quad \max_{1 \le i \le m} |f_i(x)|.$$

While these approaches are sometimes used in practice, they both have the immediate disadvantage of involving the minimization of non-smooth functions.

As a simple example, suppose a variable z is measured at three times, i.e.

$$\text{at } t = 0, \;\; z = 0.3; \quad \text{at } t = 1, \;\; z = 0.34; \quad \text{at } t = 2, \;\; z = 0.36.$$

Using a straight line

$$z = \phi(x,t) = x_1 + x_2 t$$

as the model function we can obtain a least-squares fit to the data as follows. If $(t_i,\ z_i)$ denote the given values then the function to be minimized is

$$F(x) = \sum_{i=1}^{3} (z_i - \phi(x,t_i))^2.$$

Subsituting for the t_i and z_i we get

$$F(x) = (0.3 - x_1)^2 + (0.34 - x_1 - x_2)^2 + (0.36 - x_1 - 2x_2)^2$$

and, by differentiating,

$$\frac{\partial F}{\partial x_1} = -2(0.3 - x_1) - 2(0.34 - x_1 - x_2) - 2(0.36 - x_1 - 2x_2)$$

$$\frac{\partial F}{\partial x_2} = -2(0.34 - x_1 - x_2) - 4(0.36 - x_1 - 2x_2).$$

At a minimum of F, both partial derivatives must be zero, and so (after some simplification) it turns out that x_1 and x_2 must satisfy

$$6x_1 + 6x_2 = 2 \quad \text{and} \quad 6x_1 + 10x_2 = 2.12.$$

These equations are linear because the model is linear in x_1 and x_2. The solution is $x_1 \approx 0.303$, $x_2 \approx 0.03$ and so the best straight-line fit to the data is

$$z = 0.303 + 0.03t.$$

Exercises

1. The least-squares error function for approximating m data points (t_i, z_i) by a model function $z = \phi(x,t)$ is

$$F(x) = \sum_{i=1}^{m}(z_i - \phi(x,t_i))^2$$

Show that the gradient of this function is

$$\nabla F(x) = 2\sum_{i=1}^{3}(z_i - \phi(x,t_i))\nabla\phi(x,t_i).$$

2. Given the data

t	0	1	2	3	4	5	6
z	0.4	0.42	0.43	0.46	0.47	0.5	0.49

determine the function F which must be minimized to give a least-squares fit using the model $z = x_1 + x_2t$. Also obtain expressions for ∇F and $\nabla^2 F$.

3. For the data in question 2 we can compute a *moving average* $\bar{z}$ for $2 \leq t \leq 6$ using the formula

$$\bar{z}_i = \frac{z_i + z_{i-1} + z_{i-2}}{3}.$$

Determine the function which must be minimized to give a least squares fit using the model $\bar{z} = x_1 + x_2t$.

2. The Gauss-Newton method

We can obtain a variant of Newton's method for the special case when the function to be minimized is a *sum of squared terms*, i.e.

$$F(x) = \sum_{i=1}^{m} f_i(x)^2 \tag{11.2.1}$$

where we *assume* $m \geq n$. Differentiating (11.2.1) we get

$$\nabla F(x) = 2\{\sum_{i=1}^{m} \nabla f_i(x) f_i(x)\}$$

If f is the m-vector whose elements are the *subfunctions* $f_i(x)$ and if J is the $m \times n$ *Jacobian* matrix whose i-th row is $\nabla f_i(x)^T$ then

$$\nabla F(x) = 2J^T f. \qquad (11.2.2)$$

Differentiating a second time gives

$$\nabla^2 F(x) = 2\{J^T J + \sum_{i=1}^{m} \nabla^2 f_i(x) f_i(x)\}. \qquad (11.2.3)$$

In data-fitting problems, the subfunctions are often close to zero at a minimum of (11.2.1). It may also happen that the model function is chosen so that the f_i are nearly linear and hence $||\nabla^2 f_i(x)||$ is close to zero. In both situations we can *ignore* the second term on the right hand side of (11.2.3). In other words, we can treat $2J^T J$ as a convenient approximation to $\nabla^2 F$. This leads to an algorithm which resembles the Newton method but uses no second derivatives.

The Gauss-Newton Method for minimizing a sum of squares

Choose x_0 as an estimate of x^*
Repeat for $k = 0, 1, 2, ...$
  Set $f_k =$ the vector with elements $f_i(x_k)$
  Set J_k as the corresponding Jacobian
  Obtain p_k by solving $(J_k^T J_k) p_k = -J_k^T f_k$
  Find s so $F(x_k + s p_k)$ satisfies Wolfe conditions 2 and 3
  Set $x_{k+1} = x_k + s p_k$
until $||J_k^T f_k||$ is sufficiently small.

The vector p_k used in this algorithm approximates the Newton direction because $2J_k^T f_k = \nabla F(x_k)$ and $2J_k^T J_k \approx \nabla^2 F(x_k)$. Since $J_k^T J_k$ can be shown to be positive semi-definite we can (fairly) safely assume that p_k is a descent direction, satisfying Wolfe condition 1. The equations which give the Gauss-Newton search direction are called the *normal equations*.

The Gauss-Newton algorithm can often minimize a function of the form (11.2.1) in fewer iterations than more general unconstrained optimization methods. However, it also does more work per iteration than, say, a quasi-Newton method since $O(n^2 m) + O(\frac{1}{6} n^3)$ multiplications are needed to form $J_k^T J_k$ and then factorize it by the Cholesky method.

In the exceptional case that $J_k^T J_k$ is singular, the Gauss-Newton algorithm given above will fail. However, if we choose some $\lambda > 0$ we can obtain a downhill search direction, p_k, from the *Levenberg- Marquardt* equations [36, 37]

$$(J_k^T J_k + \lambda I) p_k = -J_k^T f_k.$$

As in section 4 of chapter 6, this search direction minimizes a quadratic model of F subject to a limit on the Euclidian norm of p, i.e.,

$$\text{Minimize} \quad \frac{1}{2}(p^T J_k^T J_k p) + p^T J_k^T f_k \quad \text{subject to} \quad ||p||_2 \leq \Delta$$

for some positive Δ. The relationship between λ and Δ is not simple, but we can easily see that, as $\lambda \to \infty$, p_k tends towards an infinitesimal step along the steepest descent direction $-J_k^T f_k$.

Exercises

1. Prove that the matrix $J_k^T J_k$, used in the Gauss-Newton algorithm, is positive semi-definite.

2. Solve problems 2 and 3 from section 1 using the Gauss-Newton algorithm.

3. Least-squares in time series analysis

Suppose we have a sequence of observed values z_i, $i = 1,..,m$ at times $t_1,..,t_m$. These might be daily prices of some security or the level of a market index such as the FTSE100. Typically, such data looks very erratic (see Figure 11.1) and is unlikely to be modelled closely by a smooth function such as a polynomial or even a periodic function such as (11.1.1).

Although it will not usually be a good fit to the data, the least-squares line is often calculated to determine the *trend* of the sequence. This will involve finding x_1 and x_2 to minimize

$$F(x) = \sum_{i=1}^{m} (z_i - x_1 - x_2 t_i)^2. \tag{11.3.1}$$

The Jacobian matrix associated with this function has m rows and 2 columns and is given by

$$J_{i1} = -1.0; \quad J_{i2} = -t_i \quad \text{for } i = 1,..,m.$$

Now suppose that x_1^* and x_2^* minimize (11.3.1) and consider the sequence of values w_i given by

$$w_i = z_i - x_1^* - x_2^* t_i. \tag{11.3.2}$$

This represents the part of the original time-series that is not modelled by the trend-line. It can be called the *de-trended* data. In order to try and fit a pattern to the time-series for w we can use an *autoregressive* model [38]

$$w_i = \xi_1 w_{i-1} + \xi_2 w_{i-2} + \xi_3 \tag{11.3.3}$$

where the ξ_i are coefficients to be determined. (Obviously we could use a model which expresses w_i in terms of more than two previous w-values, just as we could have used a higher-order polynomial to find a trend *curve* for z.) To obtain a least squares fit by the model (11.3.3) we minimize

$$F(\xi) = \sum_{i=3}^{m} (w_i - \xi_1 w_{i-1} - \xi_2 w_{i-2} - \xi_3)^2. \qquad (11.3.4)$$

The corresponding Jacobian has $m-2$ rows and 2 columns with elements

$$J_{i-2,1} = -w_{i-1}; \quad J_{i-2,2} = -w_{i-2}, \quad J_{i-2,3} = -1 \quad \text{for } i = 1,..,m.$$

We can apply the Gauss-Newton method to problems (11.3.1) and (11.3.4).

4. Gauss-Newton applied to time series

The `SAMPO` program `sample5` can be used to minimize (11.3.1) by either the Gauss-Newton or the quasi-Newton method. (A weak line search is used in both cases and so we denote the Gauss-Newton algorithm by `GNw`.) As an example we consider the following sequence of closing prices of a certain share on successive trading days

$$z = (3.85,\ 4.025,\ 4.02,\ 4.18,\ 4.09,\ 4.3,\ 4.18,\ 4.24,\ 4.13,\ 4.245). \qquad (11.4.1)$$

The times t_i are given by $t_i = i,\ i = 1,..,10$. (In practice, of course, we would normally deal with longer time series than this.) Starting from $x_1 = x_2 = 0$, `GNw` minimizes (11.3.1) in one iteration and gives the least-squares line as

$$z = 3.94 + 0.0339t. \qquad (11.4.2)$$

`QNw` gets the same result in two iterations.

Now consider (11.3.4) with w_i defined by (11.3.2) and $x_1^* = 3.94$, $x_2^* = 0.0339$. `GNw` finds the minimum of (11.3.4) at

$$\xi_1 \approx -0.0368, \quad \xi_2 \approx 0.439, \quad \xi_3 \approx 0.00575.$$

Once again, this only takes one iteration while `QNw` uses four iterations. Using definition (11.3.2) for w_i, we get the following model for z

$$z_i = 3.94575 + 0.0339t_i - 0.0368w_{i-1} + 0.439w_{i-2} \qquad (11.4.3)$$

where $w_j = z_j - 3.94 - 0.0339t_j$.

Over the range $t_3 \leq t \leq t_{10}$, (11.4.3) is a better representation of z than (11.4.2). The straight line fit (11.4.2) gives the sum of squared errors

$$\sum_{i=3}^{10}(z_i - 3.94 - 0.0339t_i)^2 \approx 0.0517$$

while for model (11.4.3) the corresponding sum of squared errors is about 0.04. Figure 11.1 compares the least-squares trend line (11.4.2) with the original data while Figure 11.2 shows the model (11.4.3).

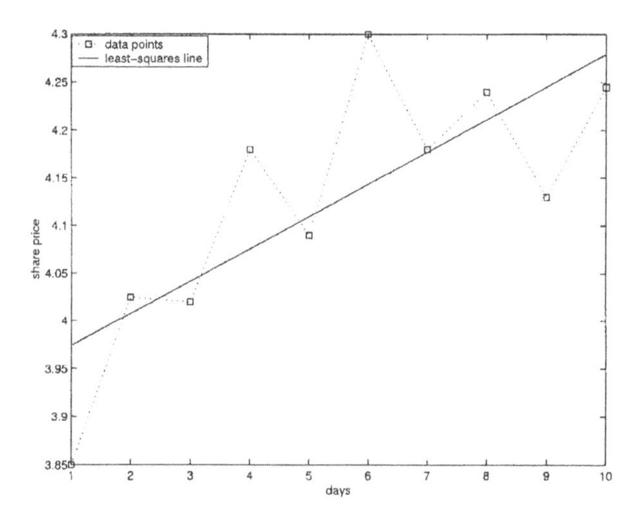

Figure 11.1. Least-squares trend line for share prices (11.4.1)

The least-squares fit by model (11.4.3) has been obtained by a two-stage process. We first find x_1^*, x_2^* to minimize (11.3.1) and then get ξ_1, ξ_2, ξ_3 by minimizing (11.3.4) which depends on x_1^* and x_2^*. Let us now use a single-stage approach to fit coefficients to the model

$$z_i = x_1 - x_2t_i - x_3(z_{i-1} - x_1 - x_2t_{i-1}) - x_4(z_{i-2} - x_1 - x_2t_{i-2})$$

which is of the same *form* as (11.4.3). This involves minimizing

$$\sum_{i=3}^{10}[z_i - x_1 - x_2t_i - x_3(z_{i-1} - x_1 - x_2t_{i-1}) - x_4(z_{i-2} - x_1 - x_2t_{i-2})]^2 \quad (11.4.4)$$

Unlike (11.3.1) and (11.3.4), this is a *non-linear* least-squares problem because the subfunctions in (11.4.4) are not linear in $x_1, .., x_4$. Hence, when we use the Gauss-Newton method to minimize (11.4.4) we do not expect it to converge in one iteration. In fact, `GNw` needs only three iterations when started from the

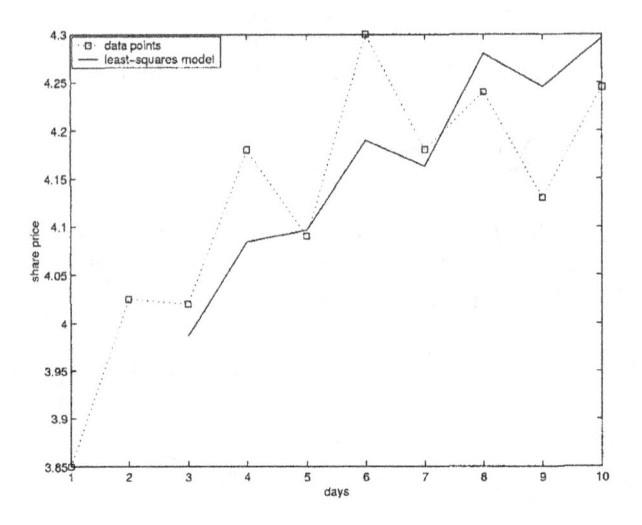

Figure 11.2. Least-squares model (11.4.3) for share prices (11.4.1)

guessed solution $x = (3.94,\ 0.0339,\ -0.0368,\ 0.439)^T$. `QNw` takes 28 iterations, however, which shows that the specialised Gauss-Newton method can be much more efficient than a general-purpose minimization method on on a sums-of-squares problem. Both methods give the same solution, namely

$$x_1 \approx 4.83, \quad x_2 \approx -0.0494, \quad x_3 \approx -0.164, \quad x_4 \approx 0.812. \qquad (11.4.5)$$

Putting these values in (11.4.4) leads to a significantly better model than (11.4.3). The sum of squared errors from t_3 to t_{10} is about 0.013 – that is, about one-third of the least-squares error function obtained with (11.4.3). Figure 11.3 shows that the model (11.4.4) comes much closer to the original data.

Exercises (To be solved using `sample5` or other suitable software.)

1. Write down expressions for the elements of the Jacobian matrix for the function (11.4.4).

2. Using the model function for (11.4.1) obtained by minimizing (11.4.4), calculate an extrapolated value for the share price on day 11. Given that the closing prices on days 11 to 15 are 4.165, 4.34, 4.23, 4.295, 4.37, calculate new least-squares coefficients for the model, based on data for days 1 to 14. Hence obtain an extrapolated estimate for the price on day 15. How well does it compare with the actual price?

3. Suppose we use a five parameter model for z, of the form

$$z_i = x_5 - x_2 t_i - x_3(z_{i-1} - x_1 - x_2 t_{i-1}) - x_4(z_{i-2} - x_1 - x_2 t_{i-2}).$$

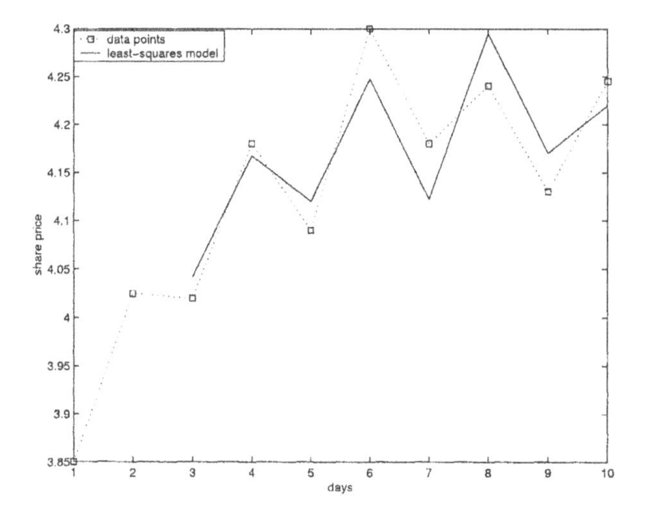

Figure 11.3. Least-squares model (11.4.4) for share prices (11.4.1)

Find the Jacobian matrix for the least-squares error function associated with this model. Use Gauss-Newton and quasi-Newton methods to minimize this error function. Comment on the results you obtain.

4. Derive a function similar to (11.4.4) but based on using a quadratic trend curve for z, rather than a trend line.

5. Least-squares forms of minimum-risk problems

In section 6 of chapter 1 we showed that the risk V can be expressed as

$$V = y^T Q y = \frac{1}{m} y^T A^T A y$$

where A is the matrix (1.6.2) whose elements are defined by $a_{ki} = (r_{ik} - \bar{r}_i)$. Since

$$Ay = \begin{pmatrix} \sum_{i=1}^{n} (r_{i1} - \bar{r}_i) y_i \\ \sum_{i=1}^{n} (r_{i2} - \bar{r}_i) y_i \\ \cdot\cdot \\ \sum_{i=1}^{n} (r_{im} - \bar{r}_i) y_i \end{pmatrix}$$

it follows that we can write

$$V = \frac{1}{m} \sum_{k=1}^{m} \{ \sum_{i=1}^{n} (r_{ik} - \bar{r}_i) y_i \}^2. \tag{11.5.1}$$

Hence V is a sum-of-squared terms. This enables us to formulate minimum-risk problems in a form suitable for the Gauss-Newton method.

Problem **Minrisk1u** (section 1 of chapter 9) involves minimizing

$$F(y) = \frac{\rho}{R_p^2}(\sum_{i=1}^{n} \bar{r}_i y_i - R_p)^2 + \rho(\sum_{i=1}^{n} y_i - 1)^2 + V. \quad (11.5.2)$$

Because of (11.5.1), $F(y)$ can be written in sum-of-squares form

$$F(y) = \sum_{k=1}^{m+2} f_k(y)^2 \quad (11.5.3)$$

where

$$f_1(y) = \frac{\sqrt{\rho}}{R_p}(\sum_{i=1}^{n} \bar{r}_i y_i - R_p); \quad f_2(y) = \sqrt{\rho}(\sum_{i=1}^{n} y_i - 1) \quad (11.5.4)$$

$$\text{and} \quad f_{k+2}(y) = \frac{1}{\sqrt{m}} \sum_{i=1}^{n} (r_{ik} - \bar{r}_i) y_i \quad \text{for } k = 1,..,m. \quad (11.5.5)$$

In order to use the Gauss-Newton method to minimize (11.5.2) we need to construct the Jacobian matrix whose elements are as follows

$$J_{1i} = \frac{\partial f_1}{\partial y_i} = \frac{\sqrt{\rho}\bar{r}_i}{R_p}; \quad J_{2i} = \frac{\partial f_2}{\partial y_i} = \sqrt{\rho} \quad \text{for } i = 1,..,n$$

$$\text{and} \quad J_{(k+2),i} = \frac{\partial f_{k+2}}{\partial y_i} = \frac{1}{\sqrt{m}}(r_{ik} - \bar{r}_i) \quad \text{for } k = 1,..,m, \ i = 1,..,n.$$

We now consider **Minrisk2u** from section 2 of chapter 9. Here the transformation $y_i = x_i^2$ is used to eliminate solutions which involve short-selling and so the function to be minimized is now

$$F(x) = \frac{\rho}{R_p^2}(\sum_{i=1}^{n} \bar{r}_i x_i^2 - R_p)^2 + \rho(\sum_{i=1}^{n} x_i^2 - 1)^2 + V. \quad (11.5.6)$$

V is given in terms of the x_i by

$$V = \sum_{i=1}^{n} x_i^2 (\sum_{k=1}^{n} \sigma_{ik} x_j^2) = \frac{1}{m} \sum_{k=1}^{m} \{\sum_{i=1}^{n} (r_{ik} - \bar{r}_i) x_i^2\}^2. \quad (11.5.7)$$

Hence $F(x)$ is a sum-of-squares function involving $m+2$ terms $f_k(x)^2$, where

$$f_1(y) = \frac{\sqrt{\rho}}{R_p}(\sum_{i=1}^{n} \bar{r}_i x_i^2 - R_p); \quad f_2(y) = \sqrt{\rho}(\sum_{i=1}^{n} x_i^2 - 1) \quad (11.5.8)$$

$$\text{and} \qquad f_{k+2}(y) = \frac{1}{\sqrt{m}} \sum_{i=1}^{n} (r_{ik} - \bar{r}_i) x_i^2 \quad \text{for } k = 1,..,m. \qquad (11.5.9)$$

The Jacobian matrix corresponding to these subfunctions is given by

$$J_{1i} = \frac{\partial f_1}{\partial x_i} = 2\frac{\sqrt{\rho}\bar{r}_i}{R_p} x_i; \quad J_{2i} = \frac{\partial f_2}{\partial x_i} = 2\sqrt{\rho} x_i \quad \text{for } i = 1,..,n \qquad (11.5.10)$$

$$J_{(k+2),i} = \frac{\partial f_{k+2}}{\partial x_i} = \frac{2}{\sqrt{m}} (r_{ik} - \bar{r}_i) x_i \quad \text{for } k = 1.,,.m, \; i = 1,..,n. \qquad (11.5.11)$$

Exercises

1. Show that the problems **Minrisk0** and **Minrisk1m** can be expressed as sums of squared terms. Obtain expressions for the elements of their Jacobian matrices.

2. Consider the functions

$$F_1(y) = \frac{1}{(\bar{r}^T y)^2} + \rho y^T Q y \text{ and } F_2(y) = \frac{y^T Q y}{(\bar{r}^T y)^2}$$

and explain why their minima might be expected to give a good balance between low risk and high return. Are such expectations always justified?
Using the relationships (3.1.5) and (3.1.6) to eliminate y_n, show that minimization of $F_1(y)$ and $F_2(y)$ can be reduced to problems similar (3.2.5). Show that these problems can be expressed in sum-of-squares form and write expressions for their subfunctions and corresponding Jacobian matrices.

6. Gauss-Newton applied to Minrisk1 & Minrisk2

The program `sample4` can use the Gauss-Newton method to solve minimum-risk problems in sum-of-squares form. Table 11.1 shows numbers of iterations and function calls needed by `GNw`, `NMw` and `QNw` on problems T1 and T5.

	T1 $\rho = 100$	T5 $\rho = 100$
`GNw`	1/2	1/2
`NMw`	1/2	1/2
`QNw`	10/18	6/10

Table 11.1. Comparing `GN`, `NM` and `QN` on problems T1 and T5

For problems posed in sum-of-squares form, `GNw` converges in just one iteration. The function (11.5.3) is quadratic in y and so its subfunctions are linear.

Hence the Gauss-Newton Hessian approximation $2J_k^TJ_k \approx \nabla^2F(x_k)$ is exact and the Gauss-Newton step is the same as the Newton step.

Now we consider problems T1a and T5a with short-selling excluded via formulation **Minrisk2u**. Table 11.2 shows the numbers of iterations and function calls needed by various methods to solve these problems.

	T1a $\rho = 100$	T5a $\rho = 100$
GNw	23/48	133/448 (see text)
NMw	73/101	32/44
QNw	211/336	50/79

Table 11.2. Comparing GN, NM and QN on problems T1a and T5a

The first column of Table 11.2 shows that GNw comfortably outperforms both NMw and QNw. As discussed earlier, (11.5.6) is a nonquadratic function and therefore it is more difficult to minimize than (11.5.2). In this case, the Gauss-Newton estimate of the Hessian matrix seems to be much more effective than the quasi-Newton updated approximation. The fact that GNw also outperforms NMw shows that it can be better to use the Hessian approximation $2J_k^TJ_k$ instead of the actual matrix $\nabla^2F(x_k)$. This is partly because the true Hessian is non-positive-definite at the standard starting guess used for this problem.

The good performance of GNw on problem T1a is offset by its poor behaviour on problem T5a. The solution of problem T5a is

$$y_1 \approx 0.78, \quad y_2 \approx 0.22, \quad y_3 = 0$$

as shown in Table 9.2. Hence, as the Gauss-Newton method approaches this solution, the partial derivatives (11.5.10) - (11.5.11) will involve $x_3 \approx 0$ which will cause the Jacobian matrix J to acquire a column of near-zeros. When J has a zero column, the Hessian approximation $2J^TJ$ will be singular and the Gauss-Newton search direction is not computable. (The same will happen when the columns of J are not linearly dependent.)

The SAMPO implementation of the Gauss-Newton method handles singularity of the Hessian approximation by increasing the diagonal terms of J^TJ, using the approach described in section 4 of chapter 6 for modifying non-positive-definite Hessian matrices in the Newton method. As mentioned at the end of section 2 of this chapter, this strategy can yield search directions like those given by the steepest descent method. This seems to be what happens for problem T5a. Convergence is rather slow and the count of iterations and function calls given in Table 9.2 corresponds to a *low accuracy* solution because the standard convergence test is not met within the prescribed limit of 500 iterations. This example shows that, while the Gauss-Newton method is *often* an efficient choice for minimizing functions that are sums of squared terms, the approximation on which it is based is not always valid and the method may need some further safeguards.

Exercises (To be solved with `sample4` or other suitable software)
1. With the data for problem T5, what is the largest value for target return for which `GNw` can be used successfully to find the minimum risk when short-selling is excluded?

2. Compare the performance of `GNw` and `NMw` on problem T1 when the target return is changed to $R_p = 1.17\%$.

3. The figures below are the daily percentage returns in the dataset `Real-20-5`.

Day	Asset1	Asset2	Asset3	Asset4	Asset5
1	0.00	-1.65	0.00	3.40	0.43
2	0.66	-0.62	-0.20	2.14	4.48
3	3.27	-0.62	0.20	-2.10	-1.02
4	0.00	1.34	0.00	0.00	-3.20
5	0.00	-1.06	0.00	-5.79	-0.43
6	0.00	0.00	0.00	0.00	-1.28
7	0.00	0.89	0.00	-0.91	2.82
8	-2.53	1.94	0.00	1.76	0.10
9	0.00	2.60	0.00	-1.50	0.84
10	-1.30	1.35	0.00	-0.61	-0.21
11	0.00	0.17	0.20	1.84	0.00
12	0.00	-0.25	2.84	-2.64	-0.21
13	-0.66	-0.83	0.00	-2.25	0.31
14	0.00	0.17	0.39	3.09	1.88
15	0.00	-0.08	0.00	-0.23	2.98
16	0.00	0.00	0.79	-2.16	-0.60
17	0.00	1.34	0.78	6.30	0.20
18	-0.66	1.33	0.00	-5.70	0.60
19	0.67	0.82	-0.39	0.08	0.10
20	0.00	0.49	0.00	0.47	2.88

Use this data to construct the matrix (1.6.2) and hence formulate (and solve) problems T2 and T2a in least-squares form.

Time series (2)

A range of mountains
between what was and will be:
I must find the pass.

Chapter 12

EQUALITY CONSTRAINED OPTIMIZATION

1. Portfolio problems with equality constraints

So far, we have dealt with portfolio selection problems in unconstrained forms such as **Minrisk1m** and **Maxret1m**. We can also express them as constrained optimization calculations. Thus a minimum-risk portfolio can be obtained by solving a problem involving *linear equality constraints*

Minrisk1

$$\text{Minimize} \quad F(y) = y^T Q y \tag{12.1.1}$$

$$\text{subject to} \quad c_1(y) = \frac{1}{R_p}(\bar{r}^T y - R_p) = 0 \tag{12.1.2}$$

$$\text{and} \quad c_2(y) = e^T y - 1 = 0 \tag{12.1.3}$$

The expected return constraint (12.1.2) includes the scaling factor $1/R_p$ for the following reason. If R_p is quite small – say $R_p = 0.1\%$ – then an absolute difference $|\bar{r}^T y - R_p| \approx 0.1$ represents a 100% error in expected return while $|e^T y - 1| \approx 0.1$ denotes only a 10% error in the condition that the invested fractions must sum to one. Convergence of numerical methods for constrained optimization is often helped if constraints are "well-balanced" and tend to zero at similar rates – i.e., if , as $y \to y^*$,

$$c_1(y) \approx c_2(y) \approx C|y - y^*|$$

for some fixed C. This is more likely to be achieved when c_1 is given by (12.1.2) instead of the more obvious $\bar{r}^T y - R_p = 0$.

We can also write a constrained version of the maximum-return problem.

Maxret1

$$\text{Minimize} \quad F(y) = -\bar{r}^T y \tag{12.1.4}$$

$$\text{subject to} \quad c_1(y) = \frac{1}{V_a}(y^T Q y - V_a) = 0 \tag{12.1.5}$$

$$\text{and} \quad c_2(y) = e^T y - 1 = 0. \tag{12.1.6}$$

The constraint (12.1.5) is expressed in scaled form for the same reasons as (12.1.2) in **Minrisk1**. Problem **Maxret1** involves a *nonlinear* equality constraint and a linear objective function. In general, nonlinearly constrained problems are harder to solve than those with linear constraints.

Both **Minrisk1** and **Maxret1** could be re-formulated using the $y_i = x_i^2$ transformation from Chapter 9 to prevent short-selling. We shall not pursue this, however, since better ways of dealing with this problem (using *inequality* constraints) will be considered in a later chapter.

Both **Minrisk1** and **Maxret1** are instances of a standard form of equality-constrained minimization or *nonlinear programming* problem, namely

$$\text{Minimize } F(x) \tag{12.1.7}$$

$$\text{subject to} \quad c_i(x) = 0 \quad i = 1,..,l. \tag{12.1.8}$$

Definition If the function (12.1.7) is quadratic and the constraints (12.1.8) are linear then we have a *Quadratic Programming* (QP) problem.
Definition If x satisfies the constraints (12.1.8) it is said to be *feasible*. Otherwise it is called *infeasible*.

2. Optimality conditions

We now consider the optimality conditions at the solution of (12.1.7), (12.1.8) when the function and constraints are differentiable.

First-order conditions

The following proposition (stated without proof) gives the optimality conditions for an equality constrained problem. (These are known as the *Karush-Kuhn-Tucker (KKT)* conditions.)

Proposition If x^* is a local solution of the equality constrained problem (12.1.7), (12.1.8) then *(i)* the point x^* must be feasible, i.e.,

$$c_i(x^*) = 0 \ (i = 1, ..., l) \tag{12.2.1}$$

and *(ii)* there must exist scalars $\lambda_1^*, ..., \lambda_l^*$ such that

$$\nabla F(x^*) - \sum_{i=1}^{l} \lambda_i^* \nabla c_i(x^*) = 0. \qquad (12.2.2)$$

Definition The quantities $\lambda_1^*, ..., \lambda_l^*$ in (12.2.2) are called *Lagrange multipliers.*

Definition The vectors $\nabla c_1(x), .., \nabla c_l(x)$ are called the constraint *normals.* For the Lagrange multipliers λ_i^* to be unique the constraint normals $\nabla c_i(x^*)$ must be linearly independent.

Definition The $l \times n$ matrix with rows $\nabla c_1(x)^T, .., \nabla c_l(x)^T$ is known as the *Jacobian* of the constraints.

If N is the Jacobian matrix of the constraints (12.1.8) then (12.2.2) implies

$$\nabla F(x^*) - N^T \lambda^* = 0. \qquad (12.2.3)$$

Conditions (12.2.2) and (12.2.3) imply that the gradient $\nabla F(x^*)$ is linearly dependent on the constraint normals. This reflects the fact that a constrained minimum represents a balance between the function and the constraints. We cannot move away from x^* without *either* violating a constraint *or* increasing the function value. This can be stated formally as follows.

Proposition If x^* solves (12.1.7), (12.1.8) and $x^* + \delta x$ is a nearby point then *(i)* if $F(x^* + \delta x) < F(x^*)$ then $c_i(x^* + \delta x) \neq 0$ for some i;
(ii) if $c_1(x^* + \delta x) = ... = c_l(x^* + \delta x) = 0$ then $F(x^* + \delta x) \geq F(x^*)$.

The left hand side of (12.2.2) can be regarded as the gradient of a function

$$L(x, \lambda^*) = F(x) - \sum_{i=1}^{l} \lambda_i^* \, c_i(x) = F(x) - \lambda^{*T} c(x). \qquad (12.2.4)$$

Definition $L(x, \lambda^*)$ is the *Lagrangian* function for problem (12.1.7), (12.1.8).

Feasible directions & second order conditions

Definition An n-vector, z, is said to be a *feasible direction* at x^* if $Nz = 0$, where N is the matrix of constraint normals appearing in (12.2.3).

Without loss of generality, we assume z is a feasible direction normalized so that $||z|| = 1$. If we consider the Taylor expansion

$$c(x^* + \varepsilon z) = c(x^*) + \varepsilon N z + O(||\varepsilon z||^2)$$

then $c(x^* + \varepsilon z) = O(\varepsilon^2)$. Therefore a move away from x^* along z keeps the constraints satisfied to first order accuracy. In particular, if all the constraints

(12.1.8) are linear then $x^* + \varepsilon z$ is a feasible point for all ε. If any of the $c_i(x)$ are nonlinear then z defines a direction *tangential* to the constraints at x^*.

Proposition Condition (12.2.3) implies that, for a feasible direction z,

$$z^T \nabla F(x^*) = 0. \tag{12.2.5}$$

Proof The result follows on premultiplying (12.2.3) by z^T.

Expressions (12.2.1) and (12.2.2) are first order conditions that hold at any constrained stationary point. To distinguish a minimum from a maximum or a saddle point we need a *second order* condition which can be stated as follows.
Proposition When the constraint functions c_i are all linear, the second order condition guaranteeing that x^* is a minimum of problem (12.1.7), (12.1.8) is

$$z^T \nabla^2 F(x^*) z > 0 \tag{12.2.6}$$

for *any* feasible direction z.

For problems with nonlinear constraints it is the Hessian of the Lagrangian function (12.2.4) which appears in the second order optimality condition.
Proposition When the constraint functions are nonlinear, the second order condition that guarantees x^* is a minimum of problem (12.1.7), (12.1.8) is

$$z^T \nabla^2 L(x^*, \lambda^*) z > 0 \tag{12.2.7}$$

for any feasible direction z.

Exercises

1. Use Taylor series arguments to show that *for a problem with linear constraints* the optimality conditions (12.2.1), (12.2.2) and (12.2.6) ensure that if δx is such that if $c_1(x^* + \delta x) = .. = c_l(x^* + \delta x) = 0$ then

$$F(x^* + \delta x) \geq F(x^*). \tag{12.2.8}$$

If the constraints are *quadratic* show that conditions (12.2.1), (12.2.2) and (12.2.7) will cause (12.2.8) to hold when $c_i(x^* + \delta x) = .. = c_l(x^* + \delta x) = 0$.

2. Show that, if G is positive definite, the problem

$$\text{Minimize } \frac{1}{2}x^T Gx + h^T x \text{ subject to } x^T x = 1$$

has a solution given by $x = -(\lambda I + G)^{-1} h$ for some scalar λ.
How does this result relate to trust region methods (chapter 6)?

3. Show that, if the constraints (12.1.8) are all divided by a constant factor k, the solution of the modified nonlinear programming problem is unchanged except that the new Lagrange multipliers are given by $k\lambda_1^*, .., k\lambda_l^*$.

4. If (12.1.7), (12.1.8) has a solution x^* where the constraint normals are not linearly independent show that $\lambda_1^*,..,\lambda_l^*$ are not uniquely defined.

3. A worked example

In some cases the optimality conditions (12.2.1), (12.2.2) can be used directly to find a solution (x^*, λ^*). Consider the problem

$$\text{Minimize } F(x) = x_1^2 + 3x_1x_2 \text{ subject to } c_1(x) = x_1 + 5x_2 - 1 = 0 \quad (12.3.1)$$

The optimality conditions mean that x_1^*, x_2^* and λ_1^* satisfy the three equations

$$x_1 + 5x_2 - 1 = 0$$

$$\frac{\partial F}{\partial x_1} - \lambda_1 \frac{\partial c_1}{\partial x_1} = 2x_1 + 3x_2 - \lambda_1 = 0$$

$$\frac{\partial F}{\partial x_2} - \lambda_1 \frac{\partial c_1}{\partial x_2} = 3x_1 - 5\lambda_1 = 0.$$

We can deduce from the last equation that $\lambda_1 = \frac{3}{5}x_1$ and then the second equation gives $x_2 = -\frac{7}{15}x_1$. Hence the first equation reduces to $-\frac{4}{3}x_1 - 1 = 0$ and so the constrained minimum occurs at

$$x_1^* = -\frac{3}{4}, \; x_2^* = \frac{7}{20} \text{ with Lagrange multiplier } \lambda_1^* = -\frac{9}{20}.$$

Exercises

1. Use the optimality conditions to solve

$$\text{Minimize} \quad x_1^2 + x_2^2 \quad \text{subject to} \quad x_1x_2 = 1.$$

2. Find a solution to the problem

$$\text{Minimize} \quad -x_1^2 - x_2^2 + x_3^2 \text{ subject to } x_1 + x_2 = -1 \text{ and } x_1^2 + x_2^2 = \frac{1}{2}$$

and comment on the values of the Lagrange multipliers.

4. Interpretation of Lagrange multipliers

The Lagrange multipliers at the solution of a constrained optimization problem are not simply mathematical abstractions. They can be used as measures of the sensitivity of the solution with respect to changes in the constraints. Suppose that x^* solves the problem

$$\text{Minimize} \quad F(x) \quad \text{subject to } c_1(x) = 0 \qquad (12.4.1)$$

and that we now consider the *perturbed* problem

$$\text{Minimize} \quad F(x) \quad \text{subject to } c_1(x) = \delta. \tag{12.4.2}$$

If the solution to (12.4.2) is $x^* + \varepsilon$ then a first order estimate of the optimum function value is

$$F(x^* + \varepsilon) \approx F(x^*) + \varepsilon \nabla F(x^*).$$

But the optimality condition for (12.4.1) states

$$\nabla F(x^*) = \lambda_1^* \nabla c_1(x^*).$$

Furthermore, since $x^* + \varepsilon$ solves (12.4.2), we must have

$$c_1(x^* + \varepsilon) = \delta$$

and so, to the first order,

$$c_1(x^*) + \varepsilon \nabla c_1(x^*) \approx \delta.$$

Putting these relationships together and using $c_1(x^*) = 0$ we get

$$F(x^* + \varepsilon) - F(x^*) \approx \delta \lambda_1^*. \tag{12.4.3}$$

Hence we have shown that the Lagrange multiplier is an approximate measure of the change in the objective function that will occur if a unit amount is added to the right hand side of the constraint. This result generalises for problems with more than one constraint. The proof of this is left to the reader.

Even though $\lambda_i^* \delta$ only gives an estimate of how much the objective function would change if the i-th constraint were shifted by δ, the Lagrange multipliers are qualitatively, as well as quantitatively, useful in practice. The Lagrange multipliers with the largest magnitude indicate which constraints have the most significant effect on the solution. In particular, in the minimum-risk problem, the Lagrange multiplier for the constraint on portfolio return will indicate the extent to which portfolio risk will be affected by an increase or decrease in the target return R_p.

Exercises

1. Extend the analysis in this section to apply to problems with more than one constraint and show that $\lambda_i^* \delta_i$ is an estimate of the change in the objective function when the i-th constraint is changed to $c_i(x) = \delta_i$.
Do a similar analysis to deal with the case when constraints i and j are shifted to become $c_i(x) = \delta_i$ and $c_j(x) = \delta_j$.

2. In the worked example of section 3, use the Lagrange multiplier to predict what the optimum function value would be if the constraint were changed to

$$x_1 + 5x_2 - \frac{4}{3} = 0.$$

Solve the problem with this modified constraint and hence determine the accuracy of the prediction. Also do similar calculations for the modified constraint

$$x_1 + 5x_2 - \frac{2}{3} = 0.$$

5. Some example problems

We now introduce test problems which will be used to illustrate the behaviour of constrained optimization methods in the following chapters.

Problem T11 is based on the data in Table 4.1 which gives expected returns (4.4.4) and variance-covariance matrix (4.4.5). It seeks the minimum-risk portfolio for a target expected return $R_p = 1\%$. The solution has invested fractions

$$y \approx (0.824,\ 0.144,\ -0.41,\ -0.672,\ 0.745) \ \text{ giving } \ V^* \approx 0.00523. \quad (12.5.1)$$

Problem T12 uses dataset `Real-20-10`. Expected returns for each asset are

$$-0.028,\ 0.366,\ 0.231,\ -0.24,\ 0.535,\ -0.17,\ -0.881,\ 0.859,\ 0.128,\ 0.087$$

and the elements of the variance-covariance matrix are

1.026	−0.434	0.020	−0.197	−0.031	−0.551	0.308	−0.093	−0.461	−0.459
−0.434	1.105	−0.078	0.235	−0.178	−0.147	−0.176	−0.45	0.177	−0.729
0.020	−0.078	0.433	−0.124	−0.189	−0.586	−0.020	−0.611	−0.209	−0.127
−0.197	0.235	−0.124	8.076	1.009	−1.879	4.558	−0.011	0.197	0.54
−0.031	−0.178	−0.189	1.009	2.901	0.027	−1.113	−0.529	−0.176	0.125
−0.551	−0.147	−0.586	−1.879	0.027	5.163	0.107	2.484	0.394	−0.142
0.308	−0.176	−0.020	4.558	−1.113	0.107	7.27	0.826	−0.26	−0.23
−0.093	−0.45	−0.611	−0.011	−0.529	2.484	0.826	5.6	0.009	1.226
−0.461	0.177	−0.209	0.197	−0.176	0.394	−0.26	0.009	0.808	0.193
−0.459	−0.729	−0.127	0.54	0.125	−0.142	−0.23	1.226	0.193	3.848

The problem is to minimize risk for an expected return $R_p = 0.1\%$. At the solution, the minimum risk is $V^* \approx 0.0335$ and the invested fractions are

$$y \approx (0.242,\ 0.146,\ -0.0015,\ 0.37,\ 0.065,\ 0.032,\ -0.015,\ 0.176,\ 0.073)$$

Problem T13 is a maximum-return problem using the same data as problem T11. The acceptable risk is $V_a = 0.0012$. The solution has invested fractions

$$y \approx (0.545,\ 0.278,\ -0.006,\ -0.074,\ 0.257) \ \text{giving}\ R^* \approx 1.105\%. \quad (12.5.2)$$

Problem T14 is a maximum-return problem, using the same data as problem T12, with $V_a = 0.025$ as the acceptable risk. The invested fractions at the solution are

$$y \approx (0.18,\ 0.164,\ 0.295,\ -0.02,\ 0.077,\ 0.016,\ 0.019,\ 0.046,\ 0.177,\ 0.045)$$

and the corresponding expected return is $R \approx 0.2159\%$.

Equality constraints

To walk a tightrope
is hard. So how much harder
to walk several?

Spiders manage it,
spinning sticky contour plots
which aren't safety nets.

Chapter 13

LINEAR EQUALITY CONSTRAINTS

1. Equality constrained quadratic programming

We now consider the special case when (12.1.7), (12.1.8) is an equality constrained QP problem (EQP), i.e.

$$\text{Minimize } \frac{1}{2}(x^T G x) + h^T x + c \quad \text{subject to } Ax + b = 0 \tag{13.1.1}$$

where the $n \times n$ matrix G and the $l \times n$ matrix A are both constant. The first order optimality conditions for (13.1.1) are

$$Ax^* + b = 0 \tag{13.1.2}$$

$$Gx^* + h - A^T \lambda^* = 0. \tag{13.1.3}$$

After re-arrangement, these become a system of $n + l$ linear equations

$$\begin{pmatrix} G & -A^T \\ -A & 0 \end{pmatrix} \begin{pmatrix} x^* \\ \lambda^* \end{pmatrix} = \begin{pmatrix} -h \\ b \end{pmatrix}. \tag{13.1.4}$$

One way of solving (13.1.1) is simply to form the linear system (13.1.4) and find x^*, λ^* using (say) Gaussian elimination. (Although (13.1.4) is symmetric it is not positive definite and so the Cholesky method is not suitable.)

If G is positive definite then the feasible stationary point obtained from (13.1.4) will necessarily be a minimum. Otherwise the second-order condition (12.2.6), with G in place of $\nabla^2 F(x^*)$, must be checked to confirm that x^* is not a constrained maximum or saddle point.

Forming and solving the whole system of equations (13.1.4) may be computationally efficient when the matrices G and A are sparse. However we can also find x^* and λ^* separately. For instance, if we multiply (13.1.3) by AG^{-1} and then use (13.1.2) to eliminate x^*, we can get λ^* from

$$(AG^{-1}A^T)\lambda^* = AG^{-1}h - b. \qquad (13.1.5)$$

It then follows that

$$Gx^* = A^T\lambda^* - h. \qquad (13.1.6)$$

The computational cost of solving (13.1.5) and (13.1.6) is $O(l^3) + O(n^3)$ multiplications which is less than the $O((n+l)^3)$ needed to solve the full system (13.1.4). However we must add the cost of forming the matrix $AG^{-1}A^T$. The inversion of G takes $O(n^3)$ multiplications (but then we can use G^{-1} to avoid the cost of solving (13.1.6) from scratch) and the matrix product takes $l^2n + ln^2$ multiplications. This fuller analysis suggests that there is little computational advantage in using (13.1.5), (13.1.6) *except* when G^{-1} is already known. In this case the solution cost is $O(l^3 + l^2n + ln^2)$ multiplications which is much less than $O((n+l)^3)$, especially when $l << n$.

Exercise
Use (13.1.5) and (13.1.6) to solve problem (12.3.1) in the previous chapter.

2. Solving minimum-risk problems as EQPs

Minrisk0

The problem **Minrisk0**

$$\text{Minimize } \; y^TQy \; \text{ subject to } \; e^Ty = 1 \qquad (13.2.1)$$

is an EQP with only one constraint. The optimality conditions (13.1.4) therefore have the rather simple form

$$2Qy^* - \lambda^* e = 0; \quad e^Ty^* = 1.$$

We have already shown in chapter 1 that the variance-covariance matrix Q can be assumed to be positive definite (and therefore nonsingular). Hence

$$y^* = \frac{\lambda^*}{2}Q^{-1}e$$

which corresponds to (13.1.6). On pre-multiplying by e^T we get

$$e^Ty^* = 1 = \frac{\lambda^*}{2}e^TQ^{-1}e.$$

It follows that the solution to **Minrisk0** has the Lagrange multiplier

$$\lambda^* = \frac{2}{e^T Q^{-1} e}.$$

Hence the optimal invested fractions are

$$y^* = (\frac{1}{e^T Q^{-1} e}) Q^{-1} e.$$

Exercise
Show that, at the solution of **Minrisk0**,

$$\lambda* = 2y^{*T} Q y^* \quad \text{and} \quad y^{*T} Q y^* = \frac{1}{e^T Q^{-1} e}.$$

Minrisk1

The problem **Minrisk1**

$$\text{Minimize} \quad y^T Q y \quad \text{subject to} \quad \frac{\bar{r}^T y - R_p}{R_p} = 0 \ \text{ and } \ e^T y - 1 = 0 \qquad (13.2.2)$$

is an EQP with two constraints. The optimality conditions (13.1.4) show that y^*, λ^* can be found by solving

$$2Qy^* - \frac{\lambda_1}{R_p}\bar{r} - \lambda_2 e = 0; \quad \frac{\bar{r}^T y - R_p}{R_p} = 0; \quad e^T y - 1 = 0.$$

It is left to the reader to show that the Lagrange multipliers for **Minrisk1** are

$$\lambda_1^* = \frac{\beta - \alpha}{\delta} \quad \lambda_2^* = \frac{\beta - \gamma}{\delta} \qquad (13.2.3)$$

where

$$\alpha = \frac{e^T Q^{-1} e}{2}, \quad \beta = \frac{\bar{r}^T Q^{-1} e}{2R_p}, \quad \gamma = \frac{\bar{r}^T Q^{-1} \bar{r}}{2R_p^2}, \quad \delta = \beta^2 - 2\alpha\gamma. \qquad (13.2.4)$$

The optimal invested fractions are then given by

$$y^* = \frac{1}{2} Q^{-1} [\frac{\lambda_1^*}{R_p}\bar{r} - \lambda_2^* e]. \qquad (13.2.5)$$

(See Steinbach [39] for further discussion.)

Exercise
Show that the solution to **Minrisk1** is given by (13.2.3) – (13.2.5).

3. Reduced-gradient methods

The essential idea behind reduced-gradient methods is the use of linear equality constraints to eliminate some variables. We have already done this with the invested fraction y_n in the portfolio problems in chapter 3, using the constraint $e^T y = 1$. More generally, suppose the linear constraints are

$$Ax + b = 0$$

and that A is partitioned as $(\tilde{A} \; : \; \bar{A})$ where $\tilde{A}$ has $n-l$ columns and $\bar{A}$ has l columns. If we also partition the vector of variables in the same way as $(\tilde{x} \; : \; \bar{x})$ then the constraints can be re-written as

$$\tilde{A}\tilde{x} + \bar{A}\bar{x} = b$$

and therefore, *if $\bar{A}$ is non-singular,*

$$\bar{x} = \bar{A}^{-1}(b - \tilde{A}\tilde{x}). \qquad (13.3.1)$$

Hence we have expressed the last l variables in terms of the first $n-l$. If the vector h and the matrix G in (13.1.1) are partitioned into $n-l$ and l columns as $(\tilde{h} \; : \; \bar{h})$ and $(\tilde{G} \; : \; \bar{G})$ the objective function can be written as

$$\frac{1}{2}(\tilde{x}^T \tilde{G}\tilde{x} + \tilde{x}^T \bar{G}\bar{x} + \bar{x}^T \tilde{G}\tilde{x} + \bar{x}^T \bar{G}\bar{x}) + \tilde{h}^T \tilde{x} + \bar{h}^T \bar{x}.$$

If we substitute for $\bar{x}$ from (13.3.1) we see that solving (13.1.1) has been reduced to finding the unconstrained minimum of an $(n-l)$-variable function.

The approach outlined in the preceding paragraphs can be implemented in a more general way that does not require us to partition the constraint matrix to find a non-singular $\bar{A}$. Instead we consider a solution in terms of its components in two subspaces which are *normal* and *tangential* to the constraints.

In chapter 12 we defined feasible directions – i.e., those lying in the tangent space of the constraints – for use in optimality conditions (12.2.5) and (12.2.6). If an n-variable optimization problem involves l linear constraints, $Ax + b = 0$, then a feasible direction can be any member of the $(n-l)$-dimensional subspace of vectors z which satisfy $Az = 0$.

Definition The $(n-l)$-dimensional subspace of vectors z satisfying $Az = 0$ is called the *null-space* of A.

We now let Z be an $n \times (n-l)$ matrix whose columns span the null-space of A. This means that, for *any* $(n-l)$-vector, w, Zw is a feasible direction. The choice of Z is not unique: but one way of obtaining it is by *orthogonal factorization* of the constraint Jacobian A. This factorization (see [17] for more details) yields an *orthonormal* $n \times n$ matrix $\bar{Q}$ and an $l \times l$ lower triangular matrix L such that

$$A\bar{Q} = R = (L:0) \quad \text{and} \quad \bar{Q}^T\bar{Q} = I. \tag{13.3.2}$$

If we now let Y be the matrix composed of the first l columns of $\bar{Q}$ and Z the matrix consisting of the remaining $(n-l)$ columns then it can be shown that

$$AZ = 0, \quad AY = L, \quad \text{and} \quad Y^TZ = 0. \tag{13.3.3}$$

Definition The l-dimensional subspace spanned by the columns of Y is called the *range-space* of the constraint Jacobian A.

Definition $Z^T\nabla F(x)$ and $Z^T\nabla^2F(x)Z$ are called, respectively, the *reduced gradient* and *reduced Hessian* of $F(x)$.

The reduced gradient approach to constrained minimization is based on considering separately the components of the solution x^* which lie in the range- and null-spaces of the constraint Jacobian. In the next subsection we show how this is done for equality constrained QP problems. Subsequently we describe the method for more general linearly constrained problems.

Reduced gradients and EQP

If x^* solves (13.1.1) we can find its components in Y- and Z-space. Suppose that $\bar{y}$ is an l-vector and $\bar{z}$ is an $(n-l)$-vector such that

$$x^* = Y\bar{y} + Z\bar{z}. \tag{13.3.4}$$

The components $Y\bar{y}$ and $Z\bar{z}$ are sometimes called the *vertical step* and the *horizontal step*. Since $AZ = 0$, optimality condition (13.1.2) implies

$$AY\bar{y} + b = 0. \tag{13.3.5}$$

This means that $\bar{y}$ can be found by solving a lower triangular system of equations. On premultiplying optimality condition (13.1.3) by Z^T we get

$$Z^TGZ\bar{z} = -Z^Tg - Z^TGY\bar{y} \tag{13.3.6}$$

which is a symmetric system of equations which can be solved for $\bar{z}$. If the EQP has a minimum then the reduced Hessian Z^TGZ is positive definite and so the Cholesky method can be applied to (13.3.6).

If we premultiply (13.1.3) by Y^T we get an upper triangular system

$$Y^TA^T\lambda^* = Y^Tg + Y^TGx^* \tag{13.3.7}$$

which can be solved to give the Lagrange multipliers.

Solving (13.3.5), (13.3.6) and (13.3.7) requires $O(l^2) + O((n-l))^3)$ multiplications which can be appreciably less than $O((n+l)^3)$ when $l \approx n$. (This comparison, however, neglects the cost of finding the Y and Z basis matrices.)

Exercises

1. Consider the problem **Minrisk0** involving two assets as stated in (13.2.1). The constraint Jacobian A is then the row-vector $(1,1)$. Show that $A\bar{Q} = (l_{11}, 0)$ when $\bar{Q}$ is a 2×2 matrix defined by

$$\bar{Q} = I - 2\frac{ww^T}{w^T w} \quad \text{with} \quad w = \begin{pmatrix} 1+\sqrt{2} \\ 1 \end{pmatrix}$$

and hence find l_{11}. Use (13.3.5) – (13.3.7) to solve **Minrisk0**.

2. Suppose that the $m \times n$ equality constraints $Ax + b = 0$ can be partitioned as

$$(A_1 \mid A_2)\begin{pmatrix} x_1 \\ x_2 \end{pmatrix} + \begin{pmatrix} b_1 \\ b_2 \end{pmatrix} = 0$$

where A_1 is $m \times m$ and non-singular. Express the EQP (13.1.1) as and unconstrained optimization calculation in terms of x_2 only.

General linearly constrained problems

The reduced gradient approach can be applied to the solution of (12.1.7), (12.1.8) when the constraints are linear but the function is non-quadratic – i.e., when the problem is of the form

$$\text{Minimize} \quad F(x) \quad \text{subject to} \quad Ax + b = 0.$$

As in unconstrained otimization, a common strategy is to use an iterative scheme based on local quadratic approximations to F. That is, in the neighbourhood of a solution estimate, x, we suppose that

$$F(x+p) \approx Q(p) = F + p^T \nabla F + \frac{1}{2}(p^T Bp) \tag{13.3.8}$$

where either $B = \nabla^2 F(x)$ or $B \approx \nabla^2 F(x)$.

One advantage of obtaining B by updating rather than by computing $B = \nabla^2 F$ is that it enables us to ensure that B is positive definite. The *BFGS* update (chapter 7) generates matrices B_{k+1} as successive estimates of $\nabla^2 F$ which satisfy the quasi-Newton condition (7.6.1). The estimates are all positive definite provided condition (7.2.2) is satisfied. Modified updating formulae have also been proposed which yield positive definite B even when $\delta_k^T \gamma_k \leq 0$. One such (called Powell's modification [40]) involves replacing the actual change in gradient, γ_k, in (7.6.2) by

$$\eta_k = (1-\theta)\gamma_k + \theta B_k \delta_k \tag{13.3.9}$$

with θ being chosen so that $\delta_k^T \eta_k > 0$.

In the unconstrained case we can justify the use of updates which force B to be positive definite because we know that $\nabla^2 F(x^*)$ must be positive definite. In constrained problems, however, the true Hessian $\nabla^2 F(x^*)$ is often indefinite. This suggests that a positive definite updating scheme may be inconsistent with making B a good approximation. In fact there is no conflict of interest because the second order optimality condition (12.2.6) only relates to the null space of the binding constraint normals. *In this subspace* the optimality conditions require the Hessian to be positive definite. For the rest of this section we shall assume that a positive definite estimate of $\nabla^2 F$ is available on every iteration.

Reduced gradient algorithm for linear equality constraints

Choose an initial *feasible* point x_0 and set $\lambda_0 = 0$
Choose B_0 as a positive definite estimate of $\nabla^2 F(x_0)$.
Set $g_0 = \nabla F(x_0)$
Obtain Y and Z as basis matrices for the range and null spaces of A
Repeat for $k = 0, 1, 2, ..$
  Determine $\bar{z}$ from $Z^T B_k Z \bar{z} = -Z^T g_k$ and set $p_k = Z\bar{z}$
  Obtain λ_{k+1} by solving $Y^T A^T \lambda = Y^T g_k + Y^T B_k p_k$
  Perform a line search to get $x_{k+1} = x_k + s p_k$ where $F(x_{k+1}) < F(x_k)$
  Set $g_{k+1} = \nabla F(x_{k+1})$
  Do a quasi-Newton update to obtain B_{k+1} from B_k
until $||Z^T g_{k+1}||$ is less than a specified tolerance.

This algorithm proceeds in a similar way to the reduced gradient method for an EQP. Each iteration makes a "horizontal" move in the subspace satisfying the constraints. No "vertical" move is needed since the algorithm described above is a *feasible point technique*. This means that it must be provided with a feasible guessed solution and then all subsequent iterates will also be feasible. (In practice, the algorithm could be preceded by some initial iterations to find a feasible point.)

A line search is needed because the non-quadraticity of F means that $x + p$ is not guaranteed to be a "better" point than x. The algorithm can use either a perfect line search to minimize $F(x + sp)$ or a weak line search to ensure that $F(x + sp) - F(x)$ is negative and bounded away from zero.

It is of course possible to implement reduced gradient methods which use exact second derivatives, but then some stages of the calculation have to be adapted to deal with indefinite matrices.

Proofs of convergence for reduced gradient algorithms can be based on ideas already discussed in relation to unconstrained minimization algorithms. Under certain, fairly mild, assumptions about the functions and constraints and about the properties of the Hessian (or its updated estimate) it can be shown that the search directions and step lengths *in the feasible subspace* will satisfy the Wolfe conditions. It then follows that the iterations will converge to a point where the reduced gradient is zero. The ultimate rate of convergence can be superlinear (or quadratic if exact second derivatives are used instead of updated approximations to the Hessian).

4. Projected gradient methods

If x_k is a feasible point for the linearly constrained problem

$$\text{Minimize } F(x) \text{ subject to } Ax+b=0.$$

then we can obtain a search direction by *projecting* a descent direction for F into the feasible subspace. For instance,

$$p_k = -(I - A^T(AA^T)^{-1}A)g_k. \tag{13.4.1}$$

is a projection of the negative gradient $-g_k$. The matrix $P = I - A^T(AA^T)^{-1}A$ is called a *projection matrix* and it is easy to show that $AP = 0$. Hence the search direction (13.4.1) satisfies $Ap_k = 0$ and a new point

$$x_{k+1} = x_k + sp_k$$

will be feasible. Hence, given an initial feasible point x_0, we can use line searches along directions given by (13.4.1) for $k = 0, 1, 2, ..$ in order to minimize F in the feasible subspace.

The projected gradient (13.4.1) is, in general, no more efficient than the steepest descent direction for unconstrained minimization. However, we can also obtain projections of more effective descent directions. If $B \approx \nabla^2 F(x_k)$ is a positive definite matrix then

$$p_k = -B^{-1}(I - A^T(AB^{-1}A^T)^{-1}A)B^{-1}g_k \tag{13.4.2}$$

is a projected quasi-Newton direction.

Clearly, projected gradient approaches work in broadly the same way as reduced gradient methods, by restricting the search to the feasible subspace. Reduced gradients have a practical advantage for large-scale problems however, because some of their computational algebra involves $n-l \times n-l$ matrices, whereas the projection methods use $n \times n$ matrices throughout.

Exercises

1. If x_k is a feasible point for EQP (13.1.1), obtain an expression for the step p such that $x_k + p$ is the solution. By writing the expression for p in the form (13.1.6), show that p can also be viewed as a projected Newton direction.

2. Write an algorithm which uses projected quasi-Newton directions to minimize $F(x)$ subject to linear constraints $Ax + b = 0$.

5. Results with methods for linear constraints

In this section we consider two methods for linearly constrained optimization. The first is an EQP method, implemented in `SAMPO` program `sample6`, which forms and solves optimality conditions (13.1.4), assuming the Hessian matrix of the objective function is explicitly available. The second is a reduced-gradient method, known as GRG (see Lasdon et. al. [41]) which is implemented in the `SOLVER` tool in Microsoft Excel [7, 8]. The GRG algorithm is broadly similar to the one in section 3 above. The `SOLVER` implementation uses *approximate* derivative information and so it will not usually deal with an EQP in a single iteration because the Hessian matrix is not available for use in equations (13.1.4). However, `SOLVER` *can* be used on problems where the function is not quadratic and – unlike the algorithm in section 3 above – it can be started from an infeasible point.

Results with `sample6` EQP

Program `sample6` can be used to find minimum-risk solutions from (13.2.2). If we solve problem T1 in this way `sample6` gives the invested fractions

$$y_1 \approx 0.421, \quad y_2 \approx 0.3372, \quad y_3 \approx 0.0094, \quad y_4 \approx 0.1935, \quad y_5 \approx 0.0389. \quad (13.5.1)$$

The corresponding risk value is 3.446×10^{-3}. The Lagrange multipliers associated with the constraints in (13.2.2) are $\lambda_1^* \approx 0.08682$ and $\lambda_2^* \approx -0.07993$.

We can compare invested fractions (13.5.1) with those obtained when we solve problem T1 via the *unconstrained* formulation **Minrisk1m** with $\rho = 100$, i.e.

$$y_1 \approx 0.4227, \quad y_2 \approx 0.3368, \quad y_3 \approx 0.0093, \quad y_4 \approx 0.19, \quad y_5 \approx 0.0413 \quad (13.5.2)$$

which gives a risk value of 3.409×10^{-3}. Clearly the two investment strategies are quite similar. However (13.5.1) is obtained with an approach which seeks to satisfy the constraint on expected return *precisely* whereas (13.5.2) comes from an unconstrained formulation which only *approximates* this requirement. Therefore the portfolio (13.5.1) can be regarded as the better solution, even though it corresponds to a slightly higher risk.

Now consider the Lagrange multiplier λ_1^* at the solution to problem T1. Since we have posed the first constraint in *scaled* form in (13.2.2) it follows that the Lagrange multiplier for the unscaled constraint $\bar{r}^T y - R_p = 0$ would be

$$\tilde{\lambda}_1^* = \frac{\lambda_1^*}{R_p} = \frac{0.08682}{1.15} \approx 0.0755.$$

Now, from section 4 of chapter 12, we know that a change in expected return from R_p to $R_p + \delta$ will change the minimum risk by about $\tilde{\lambda}_1^* \delta$. Hence we predict the minimum risk for a return $R_p = 1.16\% = 1.15 + 0.01\%$ will be

$$V \approx 0.003446 + 0.000755 = 0.0042.$$

In fact, with $R_p = 1.16\%$ the actual minimum-risk solution for problem T1 has $V \approx 0.00426$. Hence the Lagrange multiplier prediction is not exact but does give a reasonable guide to the solution of a perturbed problem.

Exercises (To be solved using `sample6` or other suitable software.)
1. Solve problem T1 by applying the quasi-Newton method to **Minrisk1m** and find, by trial-and-error, the value of ρ which leads to a solution agreeing to three significant figures with the one obtained by the EQP approach.

2. Solve problem T2 by treating it as an EQP. From the Lagrange multipliers at the solution, estimate the minimum risk if the target return is changed from 0.25% to 0.245%. How accurate is the estimate?

3. Consider the minimum-risk problem in the form

$$\text{Minimize } \; y^T Q y \; \text{ subject to } \; \bar{r}^T y - R_p = 0 \; \text{ and } \; e^T y = 1$$

and obtain expressions for its solution similar to those in (13.2.3) – (13.2.5). Also obtain expressions for the derivatives

$$\frac{\partial y_i^*}{\partial R_p} \quad \text{and} \quad \frac{\partial \lambda_j^*}{\partial R_p}$$

which show how the solution varies as the target expected return changes.

Results with SOLVER

We now use SOLVER to deal with problem T1. Briefly, this entails entering values in an *Excel* spreadsheet for the elements of $\bar{r}$ and Q, together with trial values for the invested fractions y. We then set up other spreadsheet cells to evaluate $\bar{r}^T y$, $e^T y$ and $y^T Qy$. The SOLVER dialogue box allows us to nominate the $y^T Qy$ cell as the one to be minimized by adjusting the y-values; and it also allows us to set the target value 1 for cells containing $\bar{r}^T y/R_p$ and $e^T y$.

When started from the standard initial guess $y_i = 0.2$, $i = 1,..,5$, SOLVER finds the solution in five iterations. Note that the starting point is infeasible since it gives $\bar{r}^T y \neq 1.15$ even though $e^T y = 1$. SOLVER finds a feasible point by the end of the first iteration and all subsequent steps remain feasible. If we start SOLVER from the initial guess $y_1 = .. = y_5 = 0$, which is infeasible with respect to both constraints, then six iterations are required for convergence and the method uses two iterations to obtain a feasible point.

Table 13.1 shows the number of iterations used by SOLVER for some other linearly constrained problems, using the standard initial guess $y_i = \frac{1}{n}$. The figure in brackets is the number of iterations needed to obtain a feasible point. (Unfortunately the software does not report the number of function evaluations used.) We note that, for these problems, the number of iterations is roughly the same as the number of variables.

T2	T5	T11	T12
5(1)	4(2)	5(1)	9(1)

Table 13.1. Iteration counts for SOLVER with linear constraints

Exercises

1. Obtain a table similar to 13.1 which shows the performance of SOLVER with the standard starting guesses $y_i = 0$, $(i = 1,..,n)$ and $y_i = 1$, $(i = 1,..,n)$.

2. Apply SOLVER to **Minrisk0**, using $\bar{r}$ and Q data for problems T1 and T2.

Optimality

Keep the rules and lose:
or win because you break them.
Checkmate or stalemate?

Chapter 14

PENALTY FUNCTION METHODS

1.. Introduction

We now turn to methods which can deal with nonlinear constraints in problem (12.1.7), (12.1.8). These are usually considered to present more difficulties than non-quadraticity in the objective function. This is largely because it is hard to ensure all iterates remain feasible. To illustrate this, we shall briefly consider the extension of the reduced gradient approach to deal with nonlinear equality constraints. The main focus of this chapter, however, is on methods which do not generate feasible points on every iteration but merely force the solution estimates x_k to approach feasibility as they converge.

Reduced gradients and nonlinear constraints

The reduced gradient method, described in section 3 of chapter 13, can be extended to deal with nonlinear constraints. The chief difficulty to be overcome is that of maintaining feasibility because a step along a horizontal search direction p does not now ensure that $c_i(x+sp) = 0$ for each constraint. Thus we need a *restoration* strategy in which a basic horizontal move is followed by a vertical step back onto the boundary of the feasible region. A first estimate of this retoration step can be obtained by defining

$$\hat{c}_i = c_i(x+sp) \quad i = 1,..,l,$$

finding y to solve $AYy = -\hat{c}$ and then setting $\hat{p} = Yy$. If the constraints (12.1.8) are near-linear then the point

$$x^+ = x + sp + \hat{p}$$

may be near-feasible and suitable for the start of a new iteration. However, when the c_i are highly nonlinear the calculation of a suitable restoration step may itself be an iterative process.

Another aspect of the reduced gradient algorithm that must be modified when dealing with nonlinear constraints concerns the Hessian matrix in the local quadratic model (13.3.8). The second order optimality condition for nonlinearly constrained problems is (12.2.7) which involves the Hessian of the *Lagrangian* rather than the objective function. This means that we must treat B as an approximation to

$$\nabla^2 L^* = \nabla^2 F - \sum_{i=1}^{l} \lambda_i^* \nabla^2 c_i$$

rather than to $\nabla^2 F$. A suitable update can be obtained by redefining γ_k in the quasi-Newton condition $B_{k+1}\delta_k = \gamma_k$ as

$$\gamma_k = \nabla L(x_{k+1}) - \nabla L(x_k)$$

where L is a local approximation to L^* based on Lagrange multiplier estimates, λ_{k_i}, determined at x_k, so that

$$\nabla L(x) = \nabla F(x) - \sum_{i=1}^{l} \lambda_{k_i} \nabla c_i(x).$$

If $\delta_k^T \gamma_k$ is not positive then we can use (13.3.9) to define η_k as a replacement for γ_k so that the *BFGS* update will make B_{k+1} positive definite.

Numerical results with SOLVER

We illustrate the practical difficulties that can occur when the reduced gradient method is used with nonlinear constraints by considering the GRG algorithm implemented in SOLVER. When SOLVER is applied to the maximum-return problem T3 in the form (12.1.4)–(12.1.6), using the standard initial guess $y_i = \frac{1}{3}$, a feasible point is found after seven iterations. However, this point is quite far from the optimum and for a further forty iterations the algorithm is forced to take very small steps to maintain feasibility with respect to the nonlinear constraint $y^T Qy/0.003 = 1$. SOLVER eventually terminates at

$$y_1 \approx 0.83, \; y_2 \approx 0.51, \; y_3 \approx -0.34$$

with an expected return 1.265%. This is a non-optimal point (although reasonably close to the solution (4.4.12) quoted in section 4 of chapter 4). This example shows that the reduced gradient approach can display slow convergence (or even non-convergence) when successive iterates are obtained by "creeping" around curved constraint boundaries.

Table 14.1 summarises the performance of SOLVER on some other nonlinearly constrained maximum-return problems. As in Table 13.1, the entries are numbers of iterations needed for convergence with a bracketed figure showing how many iterations are needed to obtain feasibility. An 'F' indicates that the method has terminated prematurely at a feasible but non-optimal point, after showing similar behaviour to that described in the previous paragraph.

T4	T6	T13	T14
11(1)	41(2)F	41(3)	53(6)F

Table 14.1. Iteration counts for SOLVER with non-linear constraints

Because the reduced-gradient approach can have difficulties in following curved constraints we now turn out attention to some methods which do not depend on maintaining feasibility on every iteration.

2. Penalty functions

We can avoid the difficulties associated with maintaining feasibility with respect to nonlinear constraints by converting (12.1.7), (12.1.8) into a sequence of unconstrained problems.

Definition A *penalty function* associated with (12.1.7), (12.1.8) is

$$P(x,\, r) = F(x) + \frac{1}{r}\sum_{i=1}^{l} c_i(x)^2 \qquad (14.2.1)$$

In (14.2.1), $r(>0)$ is called the *penalty parameter*. When x is a feasible point, $P(x,r) = F(x)$. When x is infeasible then P exceeds F by an amount proportional to the square of the constraint violations. An important property of the penalty function (14.2.1) is as follows.

Proposition Suppose that, in the problem (12.1.7), (12.1.8), $F(x)$ is bounded below for all x and that there is a unique solution x^* where the constraint normals $\nabla c_1(x^*),..,\nabla c_l(x^*)$ are linearly independent. Suppose also that ρ is positive and that, for all $r_k < \rho$ the Hessian matrix $\nabla^2 P(x, r_k)$ is positive definite for all x. Then if x_k solves the unconstrained problem

$$\text{Minimize } P(x, r_k) \qquad (14.2.2)$$

it follows that

$$x_k \to x^* \text{ as } r_k \to 0 \qquad (14.2.3)$$

and also

$$-\frac{2c_i(x_k)}{r_k} \to \lambda_i^* \text{ as } r_k \to 0. \qquad (14.2.4)$$

Proof The fact that $\nabla^2 P(x, r_k)$ is positive definite for r_k sufficiently small means that x_k is the *unique* minimum of $P(x, r_k)$ as $r_k \to 0$. We now show, by contradiction, that $c_1(x_k), .., c_l(x_k)$ all tend to zero as $r_k \to 0$.
Suppose this statement is false and that, for some positive constant ε,

$$\sum_{i=1}^{l} c_i(x_k)^2 > \varepsilon \text{ for all } r_k.$$

Then

$$P(x_k,\ r_k) > F(x_k) + \frac{1}{r_k}\varepsilon.$$

Now let F^* be the least value of $F(x)$ at a feasible point. Because x_k is the *unique* minimum of $P(x,\ r_k)$ it must be the case that

$$P(x_k, r_k) \leq F^*.$$

Therefore

$$F(x_k) + \frac{1}{r_k}\varepsilon < F^*.$$

Rearranging, we get

$$F(x_k) < F^* - \frac{1}{r_k}\varepsilon.$$

But, as $r_k \to 0$, this implies that $F(x_k)$ can be arbitrarily large and negative, which contradicts the condition that $F(x)$ is bounded below. Therefore, as $r_k \to 0$,

$$c_i(x_k) \to 0, \quad i = 1, .., l. \qquad (14.2.5)$$

At each unconstrained minimum, x_k,

$$\nabla P(x_k, r_k) = \nabla F(x_k) + \frac{1}{r_k}\sum_{i=1}^{l} 2c_i(x_k)\nabla c_i(x_k) = 0. \qquad (14.2.6)$$

If we define

$$\tilde{\lambda}_i(x_k) = -\frac{2}{r_k}c_i(x_k) \qquad (14.2.7)$$

then (14.2.6) is equivalent to

$$\nabla P(x_k, r_k) = \nabla F(x_k) - \sum_{i=1}^{l} \tilde{\lambda}_i(x_k)\nabla c_i(x_k) = 0. \qquad (14.2.8)$$

Now suppose that, as $r_k \to 0$, the limit point of the sequence $\{x_k\}$ is $\bar{x}$ and that $\bar{\lambda}_i = \tilde{\lambda}_i(\bar{x})$, for $i = 1,..,l$. Then, from (14.2.5) and (14.2.8)

$$c_i(\bar{x}) = 0, \quad i = 1,..,l \qquad (14.2.9)$$

$$\nabla F(\bar{x}) - \sum_{i=1}^{l} \bar{\lambda}_i \nabla c_i(\bar{x}) = 0. \qquad (14.2.10)$$

Hence $\bar{x}$ satisfies the optimality conditions for problem (12.1.7), (12.1.8). But the assumptions imply that the problem has a unique solution x^* and unique multipliers $\lambda_1^*,..,\lambda_l^*$. Therefore (14.2.3) and (14.2.4) must hold.

This result motivates the *Sequential Unconstrained Minimization Technique* (SUMT) outlined in the algorithm below. In fact, propositions similar to (14.2.3) can also be proved under weaker assumptions about the problem (12.1.7), (12.1.8). Hence, in practice, SUMT can usually be applied successfully without the need for a strict verification of the properties of the function and constraints. A full theoretical background to SUMT is given by Fiacco and McCormick [42].

Penalty Function SUMT (P-SUMT)

Choose an initial guessed solution x_0
Choose a penalty parameter r_1 and a constant $\beta(< 1)$
Repeat for $k = 1, 2, ...$
 starting from x_{k-1} use an iterative method to find x_k to solve (14.2.2)
 set $r_{k+1} = \beta r_k$
until $||c(x_k)||$ is sufficiently small

This algorithm is an example of an *infeasible* or *exterior-point* approach. The iterates x_k do not satisfy the constraints until convergence has occurred. The method does not directly calculate the Lagrange multipliers at the solution, but we can deduce their values using (14.2.4). (It should now be clear that problems **Minrisk1m** and **Maxret1m** represent a penalty function approach to constraints on return and risk, respectively.)

It is important to stress that it is not a good idea in practice to try and accelerate the progress of penalty function SUMT by choosing r_1 to be very small in the hope of getting x^* after only one unconstrained minimization. When r is near to zero, the second term in $P(x, r)$ will dominate the first. Hence, *when we evaluate P in finite precision arithmetic*, the contribution of the objective function may be lost in rounding error. Similarly, numerical evaluations of both ∇P and $\nabla^2 P$ are likely to be inaccurate when r is small. In particular, the Hessian matrix $\nabla^2 P$ tends to become *ill-conditioned* when $r \to 0$ because its condition number, defined as

$$\frac{\text{maximum eigenvalue of}\nabla^2 P}{\text{minimum eigenvalue of } \nabla^2 P},$$

can get arbitrarily large. As a consequence of all this, the numerical solution of the Newton equation $(\nabla^2 P)p = -\nabla P$ is very susceptible to rounding error and the resulting search directions can be inaccurate and ineffective. Similar difficulties can occur if we attempt to minimize $P(x,r)$ using quasi-Newton or conjugate gradient methods. The only way to avoid these numerical difficulties is to ensure that values of the $c_i(x)$ are already near-zero by the time we are dealing with very small values of r. We can best achieve this if we proceed as in the SUMT algorithm and obtain x_1, x_2, ... by relatively easy minimizations using moderately large values of the penalty parameter so that a good *and near-feasible* approximation to x^* is available by the time the unconstrained algorithm has to deal with r close to zero.

A worked example

We can demonstrate the penalty function approach on example (12.3.1). The penalty function associated with this problem is

$$P(x,r) = x_1^2 + 3x_1x_2 + \frac{1}{r}(x_1 + 5x_2 - 1)^2.$$

For any value of r, the minimum of $P(x,r)$ satisfies the equations

$$\frac{\partial P}{\partial x_1} = 2x_1 + 3x_2 + \frac{2}{r}(x_1 + 5x_2 - 1) = 0$$

$$\frac{\partial P}{\partial x_2} = 3x_1 + \frac{10}{r}(x_1 + 5x_2 - 1) = 0.$$

The second equation gives $(x_1 + 5x_2 - 1) = -\frac{3}{10}rx_1$ and on substitution in the first equation we get $x_2 = -\frac{7}{15}x_1$. Eliminating x_2 from the second equation gives $(9r - 40)x_1 - 30 = 0$ and so the minimum of $P(x,r)$ is at

$$x_1 = \frac{30}{(9r - 40)}, \quad x_2 = -\frac{210}{(135r - 600)}.$$

Hence, as $r \to 0$, the minima of $P(x,r)$ tend to

$$x_1^* = -\frac{3}{4}, \; x_2^* = \frac{7}{20}$$

which can be shown to solve (12.3.1) by direct use of the optimality conditions. The value of the constraint in (12.3.1) at the minimum of $P(x,r)$ is

$$c_1(x) = \frac{30}{(9r-40)} - \frac{1050}{(135r-600)} - 1 = -\frac{9r}{(9r-40)}$$

and hence

$$-\frac{2}{r}c_1(x) = \frac{18}{(9r-40)}.$$

If we let $r \to 0$ in the right hand side we can use (14.2.4) to deduce that the Lagrange multiplier $\lambda_1^* = -\frac{9}{20}$. This agrees with the result obtained directly from the optimality conditions.

Exercises

1. Use a penalty function approach to solve the problem

$$\text{Minimize} \quad x_1^3 + x_2^2 \quad \text{subject to} \quad x_2 - x_1^2 = 1.$$

2. Write down the penalty function $P(y,r)$ for **Minrisk0** and hence obtain an expression for $\hat{y}(r)$. (Hint: use the Sherman-Morrison-Woodbury formula (7.6.5).) Show that, as $r \to 0$, $\hat{y}(r)$ approaches the solution given in section 2 of chapter 13.

3. Suppose that we have obtained x_k, x_{k+1} as the unconstrained minima of $P(x,r_k)$ and $P(x,r_{k+1})$ respectively. Show how *linear extrapolation* could be used to obtain a first estimate of the minimum of $P(x,r_{k+2})$. Could we use a similar technique to predict the overall solution $x^* (= lim_{r_k \to 0} x_k)$?

3. The Augmented Lagrangian

The ill-conditioning difficulties which occur when minimizing $P(x,r)$ for small values of r have motivated the use of another form of penalty function [43].

Definition The *Augmented Lagrangian* is given by

$$M(x,v,r) = F(x) + \frac{1}{r}\sum_{i=1}^{l}(c_i(x) - \frac{r}{2}v_i)^2. \tag{14.3.1}$$

Compared with $P(x,r)$, the function M involves extra parameters $v_1,..,v_l$ and can also be written as

$$M(x,v,r) = F(x) - \sum_{i=1}^{l} v_i c_i(x) + \frac{1}{r}\sum_{i=1}^{l} c_i(x)^2 + \frac{r}{4}\sum_{i=1}^{l} v_i^2.$$

If we assume (12.1.7), (12.1.8) has a unique solution x^* where linear independence of $\nabla c_1(x^*),..,\nabla c_l(x^*)$ also implies uniqueness of the multiplier vector

λ^* then we can establish important properties of the Augmented Lagrangian.
Proposition The function (14.3.1) has a stationary point at $x = x^*$ for all values of r if the parameters v_i are chosen so that $v_i = \lambda_i^*$, $i = 1, ..., l$.
Proof Differentiating (14.3.1) we get

$$\nabla M(x, v, r) = \nabla F(x) + \frac{1}{r}\sum_{i=1}^{l} 2(c_i(x) - \frac{r}{2}v_i)\nabla c_i(x). \qquad (14.3.2)$$

and since $c_i(x^*) = 0$ for $i = 1, .., l$, it follows that

$$\nabla M(x^*, v, r) = \nabla F(x^*) - \sum_{i=1}^{l} v_i \nabla c_i(x^*).$$

If we set $v_i = \lambda_i^*$ $(i = 1, .., l)$ then condition (12.2.2) implies $\nabla M(x^*, \lambda^*, r) = 0$.

Proposition Suppose that ρ, σ are positive constants such that, when $r < \rho$ and $||v - \lambda|| < \sigma$, the Hessian matrix $\nabla^2 M(x, v, r)$ is positive definite for all x. Suppose also that x_k solves

$$\text{Minimize } M(x, v_k, r), \qquad (14.3.3)$$

Then, for all $r < \rho$,

$$x_k \to x^* \text{ as } v_k \to \lambda^*. \qquad (14.3.4)$$

Moreover

$$v_{k,i} - \frac{2}{r_k} c_i(x_k) \to \lambda_i^* \text{ as } x_k \to x^*. \qquad (14.3.5)$$

Proof The result (14.3.4) follows because we we have already shown that M has a stationary point at x^* when $v = \lambda^*$. The additional conditions ensure that this stationary point is a minimum. Moreover, the relationship (14.3.5) follows because $\nabla M(x_k, v_k, r_k) = 0$ and a comparison between the terms in (14.3.2) and the corresponding ones in (12.2.2) implies the required result.

Hence we can locate x^* by minimizing M when the penalty parameter r is chosen "sufficiently small". Since this is not the same as requiring r to tend to zero, it follows that we can use the penalty function (14.3.1) in a sequential unconstrained minimization technique without encountering the ill-conditioning difficulties which can occur with the function $P(x, r)$ as $r \to 0$. Such an approach needs a method of adjusting the v parameters so that they tend towards the Lagrange multipliers. A suitable technique is given in the following algorithm. The update that it uses for the parameter vector v_{k+1} is based on (14.3.5). As with the algorithm P-SUMT, this approach can still be used in practice even when the strict conditions leading to (14.3.4) cannot be verified.

Augmented Lagrangian SUMT (AL-SUMT)

Choose an initial guessed solution x_0
Choose a penalty parameter r_1 and a constant $\beta(< 1)$
Choose an initial parameter vector v_1
Repeat for $k = 1, 2, \ldots$
starting from x_{k-1} use an iterative method to find x_k to solve (14.3.4)
set $v_{k+1} = v_k - \frac{2}{r_k} c(x_k)$
set $r_{k+1} = \beta r_k$
until $||c(x_k)||$ is sufficiently small

Exercise
Obtain expressions for the gradient and Hessian of the Augmented Lagrangian function M for the equality constrained problems **Minrisk1** and **Maxret1**.

A worked example

We now demonstrate the Augmented Lagrangian approach on example (12.3.1). For this problem,

$$M(x, v, r) = x_1^2 + 3x_1x_2 - v(x_1 + 5x_2 - 1) + \frac{1}{r}(x_1 + 5x_2 - 1)^2.$$

For any value of r, the minimum of $M(x, v, r)$ satisfies the equations

$$\frac{\partial M}{\partial x_1} = 2x_1 + 3x_2 - v + \frac{2}{r}(x_1 + 5x_2 - 1) = 0 \qquad (14.3.6)$$

$$\frac{\partial M}{\partial x_2} = 3x_1 - 5v + \frac{10}{r}(x_1 + 5x_2 - 1) = 0. \qquad (14.3.7)$$

If we take $v = 0$ and $r = 0.1$ as our initial parameter choices then we can solve (14.3.6), (14.3.7) and show that the minimum of $M(x, 0, 0.1)$ occurs at

$$x_1 \approx -0.7675, \quad x_2 \approx 0.3581.$$

The value of the constraint at this point is approximately 0.023 and, by (14.3.5), the next trial value for v is

$$v \approx 0 - \frac{2}{0.1}(0.023) = -0.46$$

With this value of v (and still with $r = 0.1$) equations (14.3.6), (14.3.7) become

$$22x_1 + 103x_2 = 20 + v = 19.54$$

$$103x_1 + 500x_2 = 100 + 5v = 97.7$$

These yield $x_1 \approx -0.7495$ and $x_2 \approx 0.3498$ and so $c \approx -0.0005$. The new value of v is

$$v = -0.46 - \frac{2}{0.1}(-0.0005) \approx -0.45$$

We can see that the method is giving x_1, x_2 and v as improving approximations to the solution values of (12.3.1), namely

$$x_1^* \approx -0.75, \quad x_2^* \approx 0.35, \quad \lambda^* \approx -0.45.$$

Exercises

1. Repeat the solution of the worked example above, but using -0.5 as the initial guess for the v-parameter in the Augmented lagrangian.

2. Apply the Augmented Lagrangian method to the problem

$$\text{Minimize} \quad F(x) = x_1^2 - 4x_1x_2 + 4x_2^2 \ \text{ subject to } \ x_1 + 3x_2 + 1 = 0.$$

4. Results with P-SUMT and AL-SUMT

P-SUMT and AL-SUMT are SAMPO implementations of the sequential unconstrained minimization techniques based on $P(x, r)$ and $M(x, v, r)$. The unconstrained minimizations can be done with either QNp or QNw. Program sample6 can use these procedures to solve **Minrisk1** or **Maxret1** with initial guess (4.4.1) for the invested fractions.

Table 14.2 shows the results for problem T4 with the initial penalty parameter $r_1 = 0.1$ and the rate of decrease of the penalty parameter given by $\beta = 0.25$. For AL-SUMT the initial v-parameter vector is taken as $v_1 = 0$.

	P-SUMT			AL-SUMT		
k	$F(x_k)$	$\|\|c(x_k)\|\|$	QNw cost	$F(x_k)$	$\|\|c(x_k)\|\|$	QNw cost
1	−0.272	1.5×10^{-2}	30/46	−0.272	1.5×10^{-2}	30/46
2	−0.269	4.5×10^{-3}	39/63	−0.267	1.0×10^{-3}	39/63
3	−0.267	1.2×10^{-3}	48/79	−0.267	2.5×10^{-5}	48/79
4	−0.267	3.0×10^{-4}	55/96	−0.267	9.5×10^{-8}	53/93
5	−0.267	7.6×10^{-5}	64/115			
6	−0.267	1.9×10^{-5}	70/131			
7	−0.267	4.8×10^{-6}	75/147			

Table 14.2. P-SUMT and AL-SUMT solutions to problem T4

For each SUMT iteration, k, Table 14.2 shows the values of the objective function $F(x_k)$ (i.e. the negative of expected portfolio return), the constraint norm

$||c(x_k)||$ and the *cumulative* numbers of QNw iterations and function calls used for the unconstrained minimizations so far. We can see how successive unconstrained minima converge towards the constrained solution. Note that, for P-SUMT, the rate of reduction of the constraint norm is approximately the same as the scaling factor β. AL-SUMT, however, gives a more rapid decrease in $||c||$ on each unconstrained minimization. This is because the adjustment of the v-parameters speeds up convergence of AL-SUMT, whereas P-SUMT depends only on the reduction of r to drive the iterates x_k towards the constrained optimum.

Table 14.3 shows how performance of P-SUMT is affected by the choice of initial penalty parameter. In each case the scaling factor $\beta = 0.25$. These results, for the non-quadratic maximum-return problem T4, show that P-SUMT becomes appreciably less efficient as smaller values of r_1 are used. This confirms the comments made in section 2 that it is better to start with a moderately large value of the penalty parameter in order to ensure that we have near-feasible and near-optimal starting points for the minimizations of $P(x, r)$ when r is very small.

	P-SUMT iterations	QNw cost
$r_1 = 10^{-1}$	7	75/147
$r_1 = 10^{-2}$	5	75/138
$r_1 = 10^{-3}$	4	114/192
$r_1 = 10^{-4}$	2	151/263

Table 14.3. P-SUMT solutions to problem T4 for varying r_1

It is worth pointing out that P-SUMT *fails* on problem T4 when $r_1 = 10^{-5}$. In this case, the penalty on constraint violation is so severe that the first quasi-Newton minimization terminates at a feasible but non-optimal point.

Table 14.4 summarises the performance of P-SUMT and AL-SUMT on the four test problems T11 – T14 (section 5 of chapter 12). It shows the differences between perfect and weak line searches. (The quoted figures are for the standard settings $r_1 = 0.1$, $\beta = 0.25$ and $v_1 = 0$.)

	T11	T12	T13	T14
P-SUMT/QNw	60/100	46/114	147/240	137/280
P-SUMT/QNp	24/82	40/108	88/347	99/409
AL-SUMT/QNw	35/53	23/53	69/101	94/180
AL-SUMT/QNp	13/42	22/55	46/210	67/284

Table 14.4. Total QN iterations/function calls for P-SUMT and AL-SUMT

Table 14.4 confirms that AL-SUMT is appreciably more efficient than P-SUMT. Comparison with corresponding figures in Tables 13.1 and 14.1 shows that both the SUMT approaches take more iterations than SOLVER for the *linearly constrained* problems T11 and T12. When the constraints are non-linear, however, the SUMT approaches seem more competitive. On problem T14, unlike SOLVER, they all find the correct solution. On problem T13 AL-SUMT/QNp and SOLVER use similar numbers of iterations.

Exercises

(These problems can be solved using sample6 or other suitable software. In particular, the algorithms P-SUMT and AL-SUMT can be implemented by applying SOLVER without constraints to minimize $P(x,r)$ or $M(x,v,r)$ for a sequence of values of the parameters r and v.)

1. Use P-SUMT to solve problem T1 in the form **Minrisk1**. Since the first of the two constraints in (12.1.1) - (12.1.3) is formulated as

$$\frac{\bar{r}^T y - R_p}{R_p} = 0$$

deduce from the Lagrange multipliers at the computed solution what the multiplier would be if the constraint were simply $\bar{r}^T y - R_p = 0$. Hence estimate what the minimum risk for a target return 1.1%. Check your estimate by solving the appropriate minimum-risk problem by P-SUMT.

2. Obtain results like those in Tables 14.2 and 14.3 but using problem T1. Do you observe any behaviour that is different from that described in the section above? If so, can you explain why it occurs?

3. Use P-SUMT to solve the maximum-return problem T3. Deduce, from the Lagrange multipliers at the solution, what the maximum return would be if the target risk is 0.0025 or 0.0035. Check your predictions of maximum return by solving the problem again with $V_a = 0.0025$ and $V_a = 0.0035$.
N.B. The *first* of the two constraints is formulated as $(y^T Qy - V_a)/V_a = 0$.

4. Perform numerical tests for problem T1 to discover how the speed of convergence of AL-SUMT varies with the initial choice of r_1. Why is the performance ultimately the same as P-SUMT?

5. Perform numerical trials to estimate the choices of r_1 and β for which AL-SUMT solves Problems T11 – T14 in the smallest number of iterations. How do these results compare with the figures for SOLVER?

6. Using the results in Table 14.4, comment on the advantages and drawbacks of using a perfect linesearch in P-SUMT and AL-SUMT.

7. Investigate the performance of P-SUMT and AL-SUMT on problems T13, T14 using **Maxretl** posed with *unscaled* constraints – i.e. with

$$y^T Qy - V_a = 0 \quad \text{instead of} \quad \frac{1}{V_a}(y^T Qy - V_a) = 0.$$

5. Exact penalty functions

The approaches described so far are based on converting a constrained problem to a sequence of unconstrained ones. It is also possible to solve (12.1.7), (12.1.8) via a single unconstrained minimization. A function whose unconstrained minimum coincides with the solution to a constrained minimization problem is called an *exact penalty function.* As an example, consider

$$E(x,r) = F(x) + \frac{1}{r}\{\sum_{i=1}^{l} |c_i(x)|. \tag{14.5.1}$$

This is called the l_1 penalty function and it has a minimum at x^* for all r sufficiently small. It has no parameters requiring iterative adjustment and a solution of (12.1.7), (12.1.8) can be found by minimizing (14.5.1). In making this remark, of course, we assume that r has been chosen suitably. In fact there is a "threshold" condition ($r < 1/||\lambda^*||_\infty$); but normally this cannot be used in practice because the Lagrange multipliers will not be known in advance.

The function E has the undesirable property of being *non-smooth* since its derivatives are discontinuous across any surface for which $c_i(x) = 0$. This fact may cause difficulties for many unconstrained minimization algorithms which assume continuity of first derivatives.

For equality constrained problems there is a smooth exact penalty function,

$$E'(x,r) = F - c^T(AA^T)^{-1}Ag + \frac{1}{r}c^T c \tag{14.5.2}$$

where c is the vector of constraints $c_i(x)$, A is the Jacobian matrix whose rows are the constraint normals $\nabla c_i(x)^T$ and g is the gradient vector $\nabla F(x)$. The second term on the right of (14.5.2) includes a continuous approximation to the Lagrange multipliers. This follows because λ^* can be obtained from the Lagrangian stationarity condition, $g - A^T\lambda^* = 0$, by solving

$$(AA^T)\lambda^* = -Ag.$$

Hence E' is a form of Augmented Lagrangian function in which the multiplier estimates, λ, vary continuously instead of being adjusted at periodic intervals. The use of (14.5.2) was first proposed by Fletcher and Lill [44] and subsequent work based on the idea is summarised in [13].

Like (14.5.1), the exact penalty function E' has a practical disadvantage. The right hand side of (14.5.2) involves first derivatives of the function and constraints and so second derivatives of F and c_i have to be obtained if E' is to be minimized by a gradient method. Worse still, third derivatives will be needed if we wish to use a Newton algorithm.

Exercise

Solve problem T3 by forming and minimizing the exact penalty function (14.5.2) (e.g. by using SOLVER as an unconstrained minimization method). What is the largest value of r with which you obtain the correct solution?

Investigate what happens when SOLVER is applied to the non-smooth penalty function (14.5.1).

Chapter 15

SEQUENTIAL QUADRATIC PROGRAMMING

1. Introduction

Sequential Quadratic Programming (SQP) methods have become more popular than the SUMT approaches. There have been two strands of development in this area. One involves the use of successive QP approximations to (12.1.7), (12.1.8) based on linearisations of the c_i the constraints and a quadratic model of F. The other uses QP subproblems which seek to simplify the unconstrained minimization calculations in penalty function techniques.

In what follows we use the notation

$$g = \nabla F(x), \quad G = \nabla^2 F(x), \quad c = (c_1, ..., c_m)^T. \qquad (15.1.1)$$

We also let A denote the matrix whose i-th row is $\nabla c_i(x)^T$.

2. Quadratic/linear models

The first order optimality conditions at the solution (x^*, λ^*) of the equality constrained problem (12.1.7), (12.1.8) are

$$\nabla F(x^*) - \sum_{i=1}^{l} \lambda_i^* \nabla c_i(x^*) = 0 \quad \text{and} \quad c_i(x^*) = 0, \;\; i = 1, ..., l.$$

If x, λ are estimates of x^*, λ^*, we can introduce a measure of the corresponding error by defining

$$T(x, \lambda) = ||\nabla F(x) - \sum_{i=1}^{l} \lambda_i \nabla c_i(x)|| + \kappa ||c_i(x)|| \qquad (15.2.1)$$

where κ is a positive weighting parameter.

Now suppose that $\delta x = x^* - x$. Then δx and λ^* satisfy

$$\nabla F(x+\delta x) - \sum_{i=1}^{l} \lambda_i^* \nabla c_i(x+\delta x) = 0$$

and

$$c_i(x+\delta x) = 0, \text{ for } i = 1,..,l.$$

Using first order Taylor expansions we see that δx and λ^* *approximately* satisfy

$$\nabla F(x) + \nabla^2 F(x)\delta x - \sum_{i=1}^{l} \lambda_i^* \{\nabla c_i(x) + \nabla^2 c_i(x)\delta x\} = 0 \tag{15.2.2}$$

and

$$c_i(x) + \nabla c_i(x)^T \delta x = 0, \text{ for } i = 1,..,l. \tag{15.2.3}$$

If we define

$$\hat{G} = \nabla^2 F(x) - \sum_{i=1}^{m} \lambda_i^* \nabla^2 c_i(x) \tag{15.2.4}$$

(15.2.2), (15.2.3) simplify to

$$\hat{G}\delta x - A^T \lambda^* = -g \tag{15.2.5}$$

and

$$-A\delta x = c. \tag{15.2.6}$$

From (13.1.4) we deduce (15.2.5), (15.2.6) are optimality conditions for

$$\text{Minimize } \frac{1}{2}(\delta x^T \hat{G}\delta x) + g^T \delta x \text{ subject to } c + A\delta x = 0. \tag{15.2.7}$$

Hence δx and λ^* can be approximated by solving EQP (15.2.7). The objective function in (15.2.7) involves the gradient g, but its Hessian $\hat{G}$ includes second derivatives of the constraints and hence is an estimate of $\nabla^2 L$ rather than of the Hessian G. Thus nonlinearities in the constraints do appear in (15.2.7), even though its constraints are only linearisations of the c_i.

The EQP (15.2.7) can be used to calculate a search direction in an iterative algorithm for a general equality constrained minimization problem. The version of the algorithm outlined below uses a quasi-Newton estimate of $\hat{G}$ rather than calculating (15.2.4) from second derivatives. We shall refer to this as a *Wilson-Han-Powell* algorithm since these authors (independently) did much of the pioneering work in this area [45], [46], [40]

Wilson-Han-Powell SQP Algorithm (WHP-SQP)

Choose an initial point x_0 and an initial matrix B_0 approximating (15.2.4)
Repeat for $k = 0, 1, 2..$
Obtain p_k and λ_{k+1} by solving the QP subproblem

$$\text{Minimize} \quad \frac{1}{2}p^T B_k p + \nabla F(x_k)^T p$$

$$\text{subject to} \quad c_i(x_k) + \nabla c_i(x_k)^T p = 0 \quad i = 1, ..., l$$

Obtain a new point $x_{k+1} = x_k + s p_k$ via a line search.
Obtain B_{k+1} by a quasi-Newton update of B_k
until $T(x_{k+1}, \lambda_{k+1})$, given by (15.2.1), is sufficiently small

The line search in WHP-SQP may be based on ensuring that $P(x_{k+1}) < P(x_k)$, where P denotes some penalty function. Various choices for P have been tried. Some authors recommend the l_1 exact penalty function (14.5.1) while others use versions of the Augmented Lagrangian. The line search is important because, by forcing a reduction in a composite function involving both F and the c_i, it ensures that the new point x_{k+1} is, in a measurable sense, an improvement on x_k, thereby providing a basis for a proof of global convergence. Convergence of the WHP-SQP approach is discussed in [40]. It can be shown to be capable of superlinear convergence providing the updating strategy causes B_k to agree with the true Hessian of the Lagrangian *in the tangent space of the constraints*

The quasi-Newton update in WHP-SQP is typically performed using the *BFGS* formula (chapter 7) but based on the gradient of the Lagrangian function as discussed in section 1 of chapter 14.

Exercises

1. Perform one iteration of WHP-SQP applied to the problem

$$\text{Minimize} \quad x_2 \quad \text{subject to} \quad x_1^2 + x_2^2 = 1$$

starting from $x_1 = x_2 = \frac{1}{2}$ and using $B = \nabla^2[x_2 + \lambda(x_1^2 + x_2^2 - 1)]$ with $\lambda = 1$.

2. Perform one iteration of WHP-SQP applied to

$$\text{Minimize} \quad x_2^2 \quad \text{subject to} \quad x_1^2 + x_2^2 = 1 \quad \text{and} \quad x_1 + x_2 = 0.75$$

starting from $x_1 = x_2 = \frac{1}{2}$ and using $B = \nabla^2(x_2^2)$.

3. SQP methods based on penalty functions

In the Wilson-Han-Powell SQP algorithm there is no obvious connection between the calculation of the search direction and the penalty function used in the line search. We now derive an SQP algorithm in which the QP subproblem and the step length calculation are more closely related. In fact, the QP subproblem is based on approximating the minimum of the Augmented Lagrangian function (14.3.1). A Taylor expansion of $\nabla M(x,\ v,\ r)$ about x gives

$$\nabla M(x+\delta x, v, r) = g - A^T v + \frac{2}{r} A^T c + (\bar{G} + \frac{2}{r} A^T A)\delta x + .. \tag{15.3.1}$$

where

$$\bar{G} = \nabla^2 F(x) - \sum_{i=1}^{l} \nabla^2 c_i(x) v_i + \frac{2}{r}[\sum_{i=1}^{l} \nabla^2 c_i(x) c_i(x)]. \tag{15.3.2}$$

If $x+\delta x$ minimizes $M(x, v, r)$ then the left hand side of (15.3.1) is zero. Hence, neglecting higher order terms and re-arranging, we get

$$(\bar{G} + \frac{2}{r} A^T A)\delta x = -g + A^T v - \frac{2}{r} A^T c. \tag{15.3.3}$$

Solving (15.3.3) gives δx as the Newton step towards the minimum of $M(x, v, r)$.

When $x = x^*$ and $v = \lambda^*$ then, because $c_1(x^*) = ... = c_l(x^*) = 0$, (15.3.2) gives

$$\bar{G} = \nabla^2 F(x^*) - \sum_{i=1}^{m} \lambda_i^* \nabla^2 c_i(x^*).$$

Hence we can regard $\bar{G}$ as an approximation to $\nabla^2 L$.

If we define

$$u = v - \frac{2}{r}(A\delta x + c)$$

then this re-arranges to give

$$A\delta x = -\frac{r}{2}(u - v) - c. \tag{15.3.4}$$

Hence (15.3.3) simplifies to

$$\bar{G}\delta x - A^T u = -g. \tag{15.3.5}$$

Comparing (15.3.5), (15.3.4) with (13.1.4) we can see that δx and u are, respectively, the solution and the Lagrange multipliers of the EQP

$$\text{Minimize } \frac{1}{2}(\delta x^T \bar{G} \delta x) + g^T \delta x \text{ subject to } c + A\delta x = -\frac{r}{2}(u - v). \tag{15.3.6}$$

From (15.3.4) we get

$$c(x+\delta x) \approx c + A\delta x = -\frac{r}{2}(u - v).$$

This is a first-order estimate of constraint values at the minimum of $M(x, v, r)$. If $||\delta x||$ and $||c||$ are both small – which will be the case when x is near a solution – then $u \approx v$. Hence the constraints in (15.3.6) tend to linearisations of the actual problem constraints, *even when* $r \neq 0$. It follows that u – the Lagrange multipliers for (15.3.6) – can also be regarded as approximating the multipliers of the original problem.

Equations (15.3.4), (15.3.5) can be re-written as the symmetric system

$$\bar{G}\delta x - A^T u = -g \tag{15.3.7}$$

$$-A\delta x - \frac{r}{2}u = c - \frac{r}{2}v. \tag{15.3.8}$$

If we define

$$\delta v = -\frac{2}{r}(A\delta x + c),$$

so that $u = v + \delta v$ then we can re-write the equations for δx and u in terms of δx and δv, *i.e.*

$$\bar{G}\delta x - A^T \delta v = -g + A^T v \tag{15.3.9}$$

$$-A\delta x - \frac{r}{2}\delta v = c. \tag{15.3.10}$$

It can also be shown (Exercise 1, below) that we can obtain u and δx to satisfy (15.3.7), (15.3.8) by solving

$$(\frac{r}{2}I + A\bar{G}^{-1}A^T)u = A\bar{G}^{-1}g - c + \frac{r}{2}v \tag{15.3.11}$$

and then using

$$\delta x = \bar{G}^{-1}(A^T u - g). \tag{15.3.12}$$

We can now give an algorithm based on the preceding discussion. As with WHP-SQP, we describe a version which uses a quasi-Newton update. In this case we use an estimate of the matrix $\bar{G}^{-1}$ which approximates the inverse Hessian of the Lagrangian. If $H_k \approx G^{-1}$ at the start of iteration k then a search direction p_k and multipliers u_k are obtained by solving a QP subproblem of the same form as (15.3.6), namely

$$\text{Minimize } \frac{1}{2}p^T H_k^{-1} p + p^T \nabla F(x_k)$$

$$\text{subject to } c_i(x_k) + \nabla c_i(x_k)^T p = -\frac{r_k}{2}(u_{k_i} - \lambda_{k_i}) \text{ for } i = 1,..,l$$

where λ_k are the Lagrange multiplier estimates at the start of the iteration.

Augmented Lagrangian SQP Algorithm (AL-SQP)

Choose initial values x_0 , λ_0, r_0
Choose a matrix H_0 approximating the inverse of (15.3.2)
Choose a scaling factor $\beta < 1$. Set $\mu = 0$, $T^- = T(x, \lambda_0)$.
Repeat for $k = 0, 1, 2, ..$
 Compute $c_k = c(x_k), g_k = G(x_k)$ and $A_k = A(x_k)$.
 Obtain p_k and u_k from

$$(\frac{r_k}{2}I + A_k H_k A_k^T)u_k = A_k H_k g_k - c_k + \frac{r_k}{2}\lambda_k \tag{15.3.13}$$

$$p_k = H_k(A_k^T u_k - g_k) \tag{15.3.14}$$

 Obtain a new point $x_{k+1} = x_k + sp_k$ via a line search to give

$$M(x_{k+1}, \lambda_k, r_k) < M(x_k, \lambda_k, r_k)$$

 If $T(x_k, u_k) < T^-$ then
  set $r_{k+1} = \beta r_k$, $\lambda_{k+1} = u_k$ and $T^- = T(x_{k+1}, \lambda_{k+1})$
 otherwise
  set $r_{k+1} = r_k$ and $\lambda_{k+1} = \lambda_k$
 Obtain H_{k+1} by a quasi-Newton update of H_k
until $T(x_{k+1}, \lambda_{k+1})$ is sufficiently small

The update for H_{k+1} uses the quasi-Newton condition $H_{k+1}\gamma_k = \delta_k$ in which $\gamma_k = \nabla L(x_{k+1}) - \nabla L(x_k)$ and L is the approximate Lagrangian, given by

$$L(x) = F(x) - \sum_{i=1}^{l} \lambda_{k+1} c_i(x).$$

AL-SQP can be viewed as a way of constructing an *approximation* to the "trajectory" of Augmented Lagrangian minima. The parameters r and λ are adjusted as soon as a better estimate of an optimal point is found, rather than after a complete minimization of M. This gives a quicker approach to x^* than that offered by AL-SUMT.

SQP algorithms based on the penalty function $P(x, r)$ were first suggested by Murray [47] and Biggs [48]. The Augmented Lagrangian version AL-SQP given above was first described in [49].

AL-SQP has one significant advantage over WHP-SQP – namely that its subproblems are guaranteed to have a solution. In WHP-SQP. linearisations of nonlinear constraints may be inconsistent even when the original constraints give a well-defined feasible region.

An overview of developments of both the WHP-SQP and the AL-SQP approaches can be found in [13].

Exercises

1. Show that the solution of (15.3.7), (15.3.8) can be obtained by solving

$$(\frac{r}{2}I + A\bar{G}^{-1}A^T)u = A\bar{G}^{-1}g - c + \frac{r}{2}v$$

and then using

$$\delta x = \bar{G}^{-1}(A^T u - g)$$

Show also that these expressions are together algebraically equivalent to (15.3.3) and provide an alternative way of calculating the Newton step δx.

2. How would AL-SQP be modified if we wanted to get the search direction and Lagrange multiplier estimates on each iteration from (15.3.4) and (15.3.5) *together with a quasi-Newton approximation to* $\bar{G}$?

3. Derive an algorithm similar to AL-SQP which is based on estimating the minimum of $P(x,r)$ rather than $M(x,v,r)$.

A worked example

Consider the problem

$$\text{Minimize } F(x) = \frac{1}{2}(x_1^2 + 2x_2^2) \text{ subject to } c_1(x) = x_1 + x_2 - 1 = 0. \quad (15.3.15)$$

Note that F is quadratic and c_1 is linear and so the matrix $\bar{G}$ in (15.3.2) is simply $\nabla^2 F(x)$.

Suppose $x_1 = x_2 = 1$ is a trial solution and that we use $\lambda_1 = v$ as an initial Lagrange multiplier estimate. We show the result of an iteration of AL-SQP for *any* v and penalty parameter r.

We have $g = \nabla F = (1,2)^T$ and $c_1 = 1$. Moreover

$$\bar{G} = \nabla^2 F = \begin{pmatrix} 1 & 0 \\ 0 & 2 \end{pmatrix} \quad \text{and so} \quad H = \bar{G}^{-1} = \begin{pmatrix} 1 & 0 \\ 0 & \frac{1}{2} \end{pmatrix}.$$

The matrix A is simply (1, 1). Hence

$$AHA^T = (1,\ 1)\begin{pmatrix} 1 & 0 \\ 0 & \frac{1}{2} \end{pmatrix}\begin{pmatrix} 1 \\ 1 \end{pmatrix} = \frac{3}{2}$$

$$AHg = (1,\ 1)\begin{pmatrix} 1 & 0 \\ 0 & \frac{1}{2} \end{pmatrix}\begin{pmatrix} 1 \\ 2 \end{pmatrix} = 2.$$

Thus equation (15.3.13) becomes

$$(\frac{r}{2} + \frac{3}{2})u = 2 - 1 + \frac{r}{2}v$$

so that

$$u = \frac{2}{r+3}(1 + \frac{r}{2}v).$$

Equation (15.3.14) then gives

$$p = \begin{pmatrix} 1 & 0 \\ 0 & \frac{1}{2} \end{pmatrix}\begin{pmatrix} u-1 \\ u-2 \end{pmatrix}$$

from which we get

$$p_1 = \frac{2}{r+3}(1 + \frac{r}{2}v) - 1 = \frac{rv - r - 1}{r+3}$$

$$p_2 = \frac{1}{r+3}(1 + \frac{r}{2}v) - 1 = \frac{rv - 2r - 4}{2(r+3)}.$$

Hence, the new approximation, $x + p$, to the minimum of $M(x, v, r)$ is given by

$$x^+ = (1 + \frac{rv - r - 1}{r+3},\ 1 + \frac{rv - 2r - 4}{2(r+3)})^T.$$

For *any* value of v it is clear that, as $r \to 0$, x^+ tends to $(\frac{2}{3},\ \frac{1}{3})^T$, which solves (15.3.15). The Lagrange multiplier at the solution $x^* = (\frac{2}{3},\ \frac{1}{3})^T$ of (15.3.15) can easily be shown to be $\lambda_1^* = \frac{2}{3}$.

Now suppose that we use the parameter value $v = \lambda_1^* = \frac{2}{3}$ in the calculations above for obtaining u and p. (15.3.13) gives

$$u = \frac{2}{r+3}(1 + \frac{r}{2}v) = \frac{2}{r+3}(1 + \frac{r}{3}) = \frac{2}{3}$$

and from (15.3.14) we get

$$p_1 = -\frac{1}{3} \text{ and } p_2 = -\frac{2}{3}.$$

Hence, when $v = \lambda_1^*$, we get $x^+ = x^*$ and $u = \lambda_1^*$ *for any value of* r.

Exercises

1. Repeat the worked example from this section but calculating u and p from (15.3.7) and (15.3.8).

2. Perform one iteration of AL-SQP applied to

$$\text{Minimize} \quad F(x) = x_2^2 \quad \text{subject to} \quad x_1^2 + x_2^2 = 1 \quad \text{and} \quad x_1 + x_2 = 0.75$$

starting from $x_1 = x_2 = \frac{1}{2}$ and using $B = \nabla^2 F(x)$. How does this compare with the behaviour of WHP-SQP?

4. Results with AL-SQP

AL-SQP denotes the SAMPO implementation of AL-SQP, in which the inverse Hessian estimate H is obtained using the Powell modification to the BFGS update [40]. Note that this method is *only* implemented with a weak line search.

The program sample6 can be used to apply AL-SQP to **Minrisk1** and **Maxret1**. For Problem T4 (with $r_1 = 0.1$, $\beta = 0.25$) we obtain the convergence history shown in Table 15.1.

k	$F(x_k)$	$\|\|c(x_k)\|\|$	itns/function calls
1	−0.355	1.82×10^{0}	6/18
2	−0.376	1.05×10^{0}	8/21
3	−0.321	3.48×10^{-1}	9/22
4	−0.290	1.37×10^{-1}	10/23
5	−0.274	2.58×10^{-2}	12/25
6	−0.269	6.30×10^{-3}	15/28
7	−0.267	8.94×10^{-4}	16/29
8	−0.267	1.05×10^{-3}	17/30
9	−0.267	9.51×10^{-5}	18/31

Table 15.1. AL-SQP solution to problem T4 with $r_1 = 0.1$

Table 15.1 shows progress at the end of each iteration which produces a "sufficiently large" decrease in the Kuhn-Tucker error measure T (15.2.1). On these "outer" iterations the algorithm adjusts the values of r and the multiplier estimates λ_k. By comparing Table 15.1 with Table 14.2 we see that AL-SQP converges faster than either of the SUMT methods with the same values of r_1 and β. Progress towards the solution is much more rapid, in terms of iterations and function calls, when penalty parameter and multiplier estimates are updated frequently, rather than being changed only after an exact minimization of the Augmented Lagrangian.

Table 15.2 shows the numbers of iterations and function calls needed by AL-SQP to solve problems T11–T14. Comparing Table 15.2 with Table 14.4 confirms the advantage of the SQP approach over SUMT. Moreover, if we consider Tables 13.1 and 14.1, we see that AL-SQP does better than SOLVER on the nonlinearly constrained problems.

	T11	T12	T13	T14
AL-SQP	21/22	10/16	13/19	43/78

Table 15.2. Performance of AL-SQP on problems T11–T14

As in section 2 of chapter 10, we can use problems T7(n) and T8(n) to compare the numbers of iterations and function evaluations needed by constrained minimization when n, the number of assets, is large. The results in Table 15.3 were all obtained using weak line searches and with the standard settings $r_1 = 0.1$, $\beta = 0.25$. Obviously the behaviour would be somewhat different if other choices were made; but the figures below give a good indication of the relative efficiencies of SQP and SUMT.

	T7(100)	T7(150)	T7(200)	T8(100)	T8(150)	T8(200)
P-SUMT	169/295	201/343	232/393	240/418	254/431	325/566
AL-SUMT	92/142	112/170	130/195	128/200	157/253	192/314
AL-SQP	44/61	50/71	67/91	81/192	91/222	119/262

Table 15.3. Performance of SUMT and SQP on problems T7 and T8

Overhead costs and run-times

We now consider how the counts of iterations and function calls in Table 15.3 translate into execution times. We have already discussed the overhead costs of QNw in chapter 10. Hence, for large n, we can estimate that each quasi-Newton iteration in P-SUMT or AL-SUMT will use about $5n^2$ multiplications ($3.5n^2$ for search direction and matrix update calculations and $1.5n^2$ for function evaluations). Let us now consider the operation count for AL-SQP. It takes $mn^2 + m^2n$ multiplications to form equations (15.3.13) and a further $\frac{1}{6}m^3 + \frac{1}{2}m^2$ multiplications to solve them by Cholesky factorization. As with QNw, we can allow $3.5n^2$ multiplications for obtaining a search direction p_k from (15.3.14) and updating the inverse Hessian estimate. Finally, Table 15.3 indicates that AL-SQP uses between 1.5 and 2.5 function calls per iteration on problems T7 and T8. Therefore the total cost of an iteration of AL-SQP is about

$$(5.5 + m)n^2 + \frac{1}{6}m^3 + (\frac{1}{2} + n)m^2 \text{ multiplications.}$$

For problem T8(n) we have $m = 2$ and so the m^2 and m^3 terms are negligible. Thus we can say that

$$\frac{\text{Solution time for } \texttt{AL-SQP}}{\text{Solution time for } \texttt{P-SUMT}} \approx \frac{7.5 I_{sqp}}{5 I_{sumt}}$$

where I_{sqp} and I_{sumt} denote numbers of iterations needed for convergence. In particular, when $n = 200$, Table 15.3 implies

$$\frac{\text{Solution time for } \texttt{AL-SQP}}{\text{Solution time for } \texttt{P-SUMT}} \approx \frac{7.5 \times 119}{5 \times 325} \approx 0.55.$$

This prediction that AL-SQP will converge in about half the time needed by P-SUMT is in quite good agreement with computational experiments. Similar analysis indicates that, for problem T8(200)

$$\frac{\text{Solution time for } \texttt{AL-SQP}}{\text{Solution time for } \texttt{AL-SUMT}} \approx 0.9.$$

Thus the time advantage of AL-SQP over AL-SUMT may not be as significant as the iteration counts suggest.

There would be a substantial difference in run-times for AL-SQP on problem T8 if the search direction and Lagrange multipliers were obtained by forming and solving equations (15.3.7), (15.3.8) rather than (15.3.13) and (15.3.14). Although the two calculations are algebraically identical (if $H = \bar{G}^{-1}$), the former takes $O((m+n)^3)$ multiplications and the total arithmetic cost per iteration is about $\frac{1}{3}(m+n)^3 + 4.5n^2$ multiplications. In particular, for problems T7 and T8 with $n = 200$, the run-time for a variant of AL-SQP which uses (15.3.7), (15.3.8) could be over 50 times greater than for the version based on (15.3.13), (15.3.14). This would not be competitive with either of the SUMT methods.

The preceding remarks about the advantages of using (15.3.13), (15.3.14) are relevant for problems like **Minrisk1** and **Maxret1** where $m << n$. However, it *might* be more efficient to obtain a search direction and Lagrange multipliers from the KKT equations for problems where $m \approx n$ and where the Hessian aproximation and the Jacobian are *sparse*. Solving an $(m+n) \times (m+n)$ sparse system could involve less arithmetic than forming and solving the dense $m \times m$ system (15.3.13) when m is relatively large.

Exercises (To be solved using `sample6` or other suitable software.)
1. Use the results in Tables 13.1, 14.1, 14.4 and 15.2 to discuss the relative performance of reduced-gradient, SUMT and SQP approaches on problems with linear and nonlinear constraints.

2. Modify problems T12 and T14 to include the constraint $y_1 + y_2 + y_3 = 0.25$ (representing a condition that a portfolio must contain a fixed contribution from

a particular sector of the market). Solve the modified problems using AL-SQP (or some other suitable method) and compare the results with those quoted in section 5 of chapter 12.

3. Transform problems T11 and T13 using the $y_i = x_i^2$ substitution to obtain solutions which exclude short-selling. Solve the transformed problems using AL-SQP (or other available software) and compare the results with those for the unmodified problems given in section 5 of chapter 12.

5. SQP line searches and the Maratos effect

We have already mentioned that a number of different penalty functions can be used as the basis of the line search in WHP-SQP. For problems with highly non-linear constraints, penalty function line searches can experience a difficulty which can be explained by considering a problem with just one constraint $c(x) = 0$. Suppose that x_k is an estimate of the solution and that p is the search direction given by the SQP subproblem at x_k. If the constraint is nonlinear and if $c(x_k)$ is quite close to zero then it is possible that

$$||x_k + p - x^*|| < ||x_k - x^*|| \quad \text{and also} \quad |c(x_k + p)| > |c(x_k)|.$$

Under these circumstances it would be appropriate to accept the new point $x_k + p$. However, suppose the line search is based on the exact penalty function

$$E(x,r) = F(x) + \frac{1}{r}|c(x)|.$$

For small values of r, the increase in constraint violation may imply

$$E(x_k + p, r) > E(x_k, r)$$

and so the line search will reject $x_k + p$ and may yield $x_{k+1} = x_k + sp$ where $s << 1$. This phenomenon is called the *Maratos effect* [50]. It can sometimes cause vary slow convergence of SQP methods when the iterates are close to the constraints (especially when near the solution).

The situation just outlined can arise whatever penalty function is used in the SQP line search. All penalty functions involve a weighted combination of the objective function and the constraint violations and the Maratos effect can occur whenever the constraint contribution is over-emphasised. Unfortunately there are no hard-and-fast rules for choosing penalty parameters so that the function and constraints will be well-balanced on every SQP iteration.

Trust-region variants of SQP methods have been developed to try to avoid such difficulties with line searches. In these approaches a step-length limit is included in the QP subproblem. However, the presence of this extra constraint introduces other difficulties (not considered here) and in any case it does not wholly eliminate the use of a penalty function in which weighting parameters have to be chosen.

Replacing line searches with filters

A more recent (and perhaps more promising) way of avoiding the Maratos effect is one which deals with function values and constraint violations separately, rather than combining them in a penalty function. We shall give a simple description of this approach, again using an example with a single equality constraint $c(x) = 0$.

Let x_0 be the initial guessed solution and write $F_0 = F(x_0)$, $e_0 = |c(x_0)|$. If p is the SQP search direction then we can accept the new point $x_1 = x_0 + sp$ if

$$\textit{either} \quad F_1 < F_0 \quad \textit{or} \quad e_1 < e_0.$$

If only one of these inequalities holds then x_0 *and* x_1 are included in a list of reference points called a *filter*. If *both* inequalities are satisfied then x_1 is said to *dominate* x_0 and the filter will consist only of the point x_1.

Now let us suppose that the filter consists of x_0 and x_1. Then the next SQP iteration will accept a point $x_2 = x_1 + sp$ if, for $j = 0, 1$,

$$\textit{either} \quad F_2 < F_j \quad \textit{or} \quad e_2 < e_j.$$

That is, a new point must be better (in terms of either function value or constraint violation) than *all* the points in the current filter. If this happens, the point x_2 will be added to the filter for use on the next iteration. Furthermore, if either x_0 or x_1 is dominated by x_2 it will be removed from the filter.

We can illustrate the idea by using a plot in (F, e)-space. In the diagram below, A, B, C and D are points in the filter at the start of an iteration. Any new point below or to the left of the dotted line is acceptable (e.g. P, Q and R). However, P does not dominate any of A, B, C or D and if this represented the new point then the filter for the next iteration would be defined by P, A, B, C, D. The point Q dominates A and B and if this were the outcome of the SQP step then the next filter would be Q, C, D. Finally, if R were the new point then it would dominate all of the current filter and the next iteration would only accept points to left of or below R.

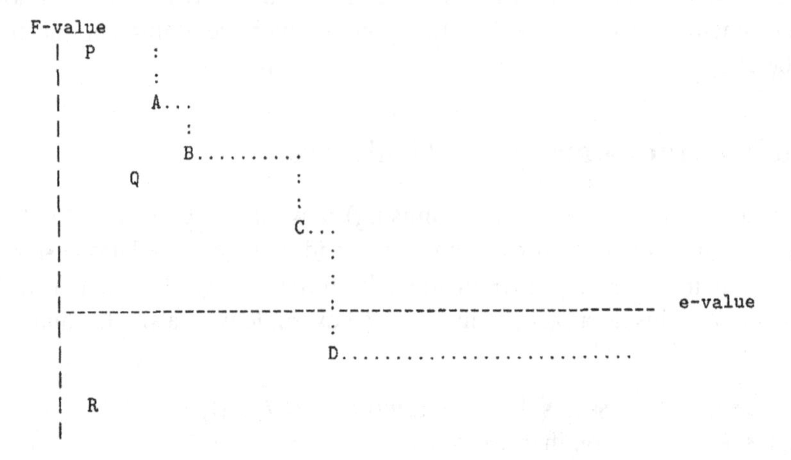

Figure 15.1. An illustration of a filter

The above description can be extended easily to problems with several constraints if we let e_k denote $||c(x_k)||$. For more details, see the original work by Fletcher and Leyffer [51].

Chapter 16

FURTHER PORTFOLIO PROBLEMS

1. Including transaction costs

We now consider versions of the minimum-risk and maximum-return problem in which *transaction costs* are taken into account. We use ideas given by Mitchell & Braun [52]. When an invested fraction y_i has been allocated to asset i, the actual purchase of that asset will normally be subject to various costs (such as commission and taxes). We can assume these costs vary linearly with the size of the purchase and so, as a fraction of the whole investment, the proportion of asset i in the final portfolio will be $(1-c_i)y_i$ for some cost factor c_i. Let us write

$$\bar{y}_i = (1-c_i)y_i \quad \text{for} \quad i = 1,..,n$$

or, in vector form,

$$\bar{y} = Dy \quad \text{where} \quad D = \mathbf{diag}(1-c_1,\ 1-c_2, ..., 1-c_n). \tag{16.1.1}$$

It is straightforward to obtain the expected portfolio return as

$$\bar{R} = \sum_{i=1}^{n} \bar{r}_i \bar{y}_i = \bar{r}^T Dy.$$

However the calculation of portfolio risk when the invested fractions are $\bar{y}_1, .., \bar{y}_n$ is slightly different from what we have seen before. The basic definition of risk as $V = y^T Qy$ is based on the standard normalisation $e^T y = 1$. When the actual investment fractions are given by $\bar{y}_i$ then $e^T \bar{y} < 1$ and we define risk by

$$\bar{V} = \frac{\bar{y}^T Q \bar{y}}{(e^T \bar{y})^2} = \frac{y^T DQDy}{(e^T Dy)^2}. \tag{16.1.2}$$

If we did not divide by $(e^T\bar{y})^2$ in the definition of $\bar{V}$ then we could trivially make risk appear small merely by choosing small invested fractions $\bar{y}_i$.

For an expected portfolio return of $R_p\%$ we can pose a new form of the minimum-risk problem which allows for transaction costs in purchasing the assets.

Minrisk3

$$\text{Minimize} \quad F(y) = \bar{V} = \frac{y^T DQDy}{(e^T Dy)^2} \tag{16.1.3}$$

$$\text{subject to} \quad c_1(y) = \frac{1}{R_p}(\bar{r}^T Dy - R_p) = 0 \tag{16.1.4}$$

$$\text{and} \quad c_2(y) = e^T y - 1 = 0. \tag{16.1.5}$$

This version of the minimum-risk problem is no longer a QP. Although the constraints are linear, the function is non-quadratic. The constraint (16.1.4) specifies a rate of return with respect to the total amount of our investment even though only part of it has actually been used to purchase assets.

Exercises

1. Find expressions for gradients of the function and constraints in **Minrisk3**.

2. Formulate a problem **Maxret3** which solves the maximum return problem, allowing for transaction costs in the same way as in **Minrisk3**. Obtain expressions for gradients of the function and constraints in this problem.

Solutions of Minrisk3

The program `sample7` solves problem **Minrisk3** using `P-SUMT`, `AL-SUMT` or `AL-SQP`. As an example we consider the introduction of transaction costs in Problem T9, based on a 75-day history of returns on the five assets. We first assume that the cost factor $c_i = c$ is the same for all assets and Table 16.1 shows how the investment fractions vary with c.

	$c=0\%$	$c=1\%$	$c=3\%$	$c=5\%$
Asset 1	0.293	0.291	0.288	0.284
Asset 2	0.365	0.365	0.365	0.365
Asset 3	0.098	0.100	0.104	0.107
Asset 4	0.139	0.139	0.139	0.139
Asset 5	0.105	0.105	0.105	0.105
$\bar{V}^*$	0.662	0.661	0.659	0.657

Table 16.1. Problem T9 with different transaction costs

The main changes in portfolio are confined to assets 1 and 3. Some adjustments are made to the other invested fractions but only in the fourth decimal place. Asset 3 is the one with largest expected return while the expected return $\bar{r}_1$ is negative. It is therefore reasonable that increasing transaction costs should lead to a decrease in y_1 and an increase in y_3.

Table 16.2 shows what happens when there are different cost factors for different assets. It gives solutions of Problem T9 when $c_i = 1\%$ for $i = 1,3,4,5$ while c_2 varies. We choose this case because y_2 is the biggest of the invested fractions in the equal-cost results in Table 16.1.

	$c_2 = 0\%$	$c_2 = 1\%$	$c_2 = 3\%$	$c_2 = 5\%$
Asset 1	0.293	0.291	0.288	0.284
Asset 2	0.363	0.365	0.370	0.375
Asset 3	0.100	0.100	0.100	0.101
Asset 4	0.139	0.139	0.138	0.137
Asset 5	0.105	0.105	0.104	0.103
$\bar{V}^*$	0.662	0.661	0.660	0.660

Table 16.2. Problem T9 with different transaction costs on asset 2

The invested fraction y_1 decreases in the same way as in the results in Table 16.1 while changes in y_3, y_4 and y_5 are slightly larger. The value of y_2 *increases* with c_2 although we might have expected investment to be switched away from asset 2 when its associated costs become higher. The fact is, however, that the presence of asset 2 in the portfolio is important for obtaining low risk; and hence y_2 has to be made larger in order that the fraction actually purchased ($\bar{y}_2 = (1 - c_2)y_2$) remains sufficiently large.

Exercises (To be solved using `sample7` or other suitable software.)

1. In Tables 16.1 and 16.2 the value of $\bar{V}^*$ decreases as transaction costs increase. Do you observe the same trend if, for the same asset history `Real-75-5`, the target return is taken as $R_p = 0.2\%$? Given that the solution of **Minrisk0** for the dataset `Real-75-5` has $V_{min} \approx 0.635$, $R_{min} \approx 0.157\%$, can you explain your observation?

2. Numerical results from `sample7` show that the Lagrange multipliers for **Minrisk3** are such that $\lambda_1 = -\lambda_2$. Can you explain why this is the case?

3. Using **Maxret3**, a version of the maximum-return problem which allows for transaction costs, determine the effect of increasing transaction costs in the solution of Problem T10 when c_i is the same for for all assets. Comment on your results.

4. Compare the performance of P-SUMT, AL-SUMT and AL-SQP on the problems discussed in this section.

2. A re-balancing problem

In section 3 of chapter 10 we mentioned the portfolio re-balancing problem. If we have values $\hat{y}_i$ for the *current* invested fractions then – in the light of more recent data about asset returns – we may wish determine changes x_i to our existing investments to get a new minimum-risk portfolio with the same target expected return. In practice, transaction costs will be involved and these will usually be proportional to the size of the changes x_i. The re-balancing problem differs from **Minrisk3** because it involves both buying and selling and so the changes x_i may be positive or negative. Thus, the cost of a transaction has to be expressed as $c_i|x_i|$ which introduces some non-differentiability into the problem. One way of avoiding this is to formulate the re-balancing problem in terms of *inequality* constraints. This will be described in a later chapter.

We can consider a simplified form of the re-balancing problem which aims to limit the *cost* of the new portfolio by finding the "smallest" changes that will meet target values for expected return and risk. If we use the two-norm as a measure of the size of x then we obtain the problem

$$\text{Minimize} \quad F(x) = x^T x \tag{16.2.1}$$

$$\text{subject to} \quad c_1(x) = \frac{1}{R_p}(\bar{r}^T(\hat{y}+x) - R_p) = 0 \tag{16.2.2}$$

$$\text{and} \quad c_2(x) = \frac{1}{V_a}[(\hat{y}+x)^T Q(\hat{y}+x) - V_a] = 0 \tag{16.2.3}$$

$$\text{and} \quad c_3(x) = e^T(\hat{y}+x) - 1 = 0. \tag{16.2.4}$$

It is easy to calculate gradients of the function and constraints for this problem.

$$\nabla F(x) = 2x, \quad \nabla c_1(x) = \frac{\bar{r}}{R_p} \quad \text{and} \quad \nabla c_3(x) = e.$$

Since c_2 can be re-written as

$$c_2(x) = \frac{1}{V_a}(x^T Qx + 2x^T Q\hat{y} + \hat{y}^T Q\hat{y} - V_a) = 0$$

we can deduce that

$$\nabla c_2(x) = \frac{1}{V_a}(2Qx + 2Q\hat{y}).$$

Solutions of re-balancing problems

To illustrate (16.2.1)–(16.2.4) we consider Problem T9, and dataset Real-75-5. As shown in the first column of Table 16.1, the optimum portfolio (without transaction costs) has invested fractions

$$\hat{y}_1 = 0.293, \;\; \hat{y}_2 = 0.365, \;\; \hat{y}_3 = 0.098, \;\; \hat{y}_4 = 0.139, \;\; \hat{y}_5 = 0.105. \qquad (16.2.5)$$

After 45 more trading days – i.e., on the basis of dataset Real-120-5 – the calculated value for portfolio expected return is about 0.082%. This is less than the 0.1% target set by Problem T9. Moreover, the risk implied by the 120-day variance-covariance matrix is $V \approx 0.84$. We can compute a re-balanced portfolio by solving (16.2.1)–(16.2.4) with $R_p = 0.1\%$ and $V_a = 0.8$. This can be done using the SAMPO program sample8, which can apply P-SUMT, AL-SUMT or AL-SQP. From an initial guess $x_1 = .. = x_n = 0$, AL-SQP takes 7 iterations (11 function calls) to obtain the solution

$$x_1 \approx -0.021, \;\; x_2 \approx 0.027, \;\; x_3 \approx 0.035, \;\; x_4 \approx -0.010, \;\; x_5 \approx -0.031$$

for the minimal changes to the current $\hat{y}_i$. The value of $x^T x$ is about 0.0034. Thus the new portfolio is defined by

$$y_1 \approx 0.272, \;\; y_2 \approx 0.392, \;\; y_3 \approx 0.133, \;\; y_4 \approx 0.129, \;\; y_5 \approx 0.074.$$

Table 16.3 shows how the solutions of (16.2.1) – (16.2.4) change with the choice of V_a. (All other problem details are as stated above.)

	$V_a = 0.81$	$V_a = 0.8$	$V_a = 0.79$	$V_a = 0.78$
Asset 1	0.264	0.272	0.283	0.305
Asset 2	0.396	0.392	0.384	0.363
Asset 3	0.129	0.133	0.141	0.165
Asset 4	0.126	0.129	0.130	0.125
Asset 5	0.085	0.074	0.061	0.043
$x^T x$	0.0033	0.0034	0.0043	0.0086

Table 16.3. Rebalancing problem T9 with different values of V_a

The solutions in Table 16.3 are *different* from the minimum-risk solution to problem T9 using dataset Real-120-5. **Minrisk1** deals with portfolio selection rather than a re-balancing, and for problem T9 and dataset Real-120-5 we get a risk $V \approx 0.779$ from a portfolio defined by

$$y_1 \approx 0.312, \;\; y_2 \approx 0.351, \;\; y_3 \approx 0.177, \;\; y_4 \approx 0.122, \;\; y_5 \approx 0.038.$$

This gives the measure of the change in the investment fractions as

$$x^T x = (\hat{y} - y)^T (\hat{y} - y) \approx 0.0108.$$

The results in Table 16.3 show that, by making more modest adjustments to the current portfolio, we can obtain the desired target return at not too much above the minimum risk.

Exercises (To be solved using `sample8` or other suitable software.)
1. What happens if we use (16.2.1)–(16.2.4) to re-balance the 75-day solution to problem T9, setting $R_p = 0.1\%$ and $V_a = 0.77$?

2. Consider the 75-day solution to problem T9 given in the first column of Table 16.1 and re-balance this portfolio on the basis of dataset `Real-120-5` to obtain an expected return of 0.2% with a risk $V_a = 0.8$.

3. Compare the numbers of iterations and function calls needed by `P-SUMT`, `AL-SUMT` and `AL-SQP` to solve the problems described in this section.

3. A sensitivity problem

Solutions to portfolio selection problems are sometimes rather "flat" – that is, there may be a region around an optimum in which changes to the invested fractions y_i do not alter the portfolio expected return or risk by very much. It may be useful to estimate how sensitive an optimized portfolio is to small variations. We can do this by formulating a problem similar to (16.2.1)–(16.2.4).

Suppose we have already obtained invested fractions $\hat{y}_1, .., \hat{y}_n$ which give a target expected return R_p with minimum risk V^*. We now wish to know the *biggest* change to the invested fractions that would still give a near-optimal performance. We might, for instance, be willing to accept an expected return $\tilde{R} = 0.99R_p$ and a risk $\tilde{V} = 1.01V^*$. To determine this we can form and solve

$$\text{Minimize} \quad F(x) = -x^T x \tag{16.3.1}$$

$$\text{subject to} \quad c_1(x) = \frac{1}{\tilde{R}}(\bar{r}^T (\hat{y} + x) - \tilde{R}) = 0 \tag{16.3.2}$$

$$\text{and} \quad c_2(x) = \frac{1}{\tilde{V}}[(\hat{y} + x)^T Q(\hat{y} + x) - \tilde{V}] = 0 \tag{16.3.3}$$

$$\text{and} \quad c_3(x) = e^T (\hat{y} + x) - 1 = 0. \tag{16.3.4}$$

Problem (16.3.1)–(16.3.4) can tell us how much "room for manoeuvre" we have in implementing an optimal solution to a portfolio selection problem. This can be useful because it may not be possible, in practice, to purchase assets in precisely the proportions indicated by the calculated invested fractions $y_1,..,y_n$. For instance, a *roundlot* requirement [53] may mean that assets have to be bought in multiples of some unit size which does not correspond exactly to the values of the y_i.

Solutions of the sensitivity problem

The results in this section were obtained with a `SAMPO` program `sample9`. Consider again the solution (16.2.5) for problem T9 with dataset `Real-75-5`. This gives $V^* \approx 0.662$. If we solve (16.3.1)–(16.3.4) with $\tilde{R} = 0.99R_p = 0.099\%$ and $\tilde{V} = 1.01V^* \approx 0.669$ we obtain the invested fractions

$$y_1 \approx 0.306,\ y_2 \approx 0.319,\ y_3 \approx 0.124,\ y_4 \approx 0.141,\ y_5 \approx 0.110.$$

Comparing this with the optimum portfolio (16.2.5) shows that quite large changes can be made to the invested fractions – particularly for assets 2 and 3 – without excessively damaging portfolio performance. If we increase the allowable margin round the optimum so that $\tilde{R} = 0.095\%$ and $\tilde{V} = 1.05V^* \approx 0.695$ then (16.3.1) – (16.3.4) yields

$$y_1 \approx 0.326,\ y_2 \approx 0.263,\ y_3 \approx 0.152,\ y_4 \approx 0.143,\ y_5 \approx 0.117.$$

A portfolio performance only about 5% less than optimal can still be obtained when y_1, y_2 and y_3 differ from their optimum values in (16.2.5) by between 10 and 20%. On the other hand, however, there seems to be much less room for manouevre in the invested fractions for assets 4 and 5.

Exercises (To be solved using `sample9` or other suitable software.)
1. Solve the maximum-return problem T10 using the dataset `Real-75-5` and then determine the biggest changes in invested fractions which will yield 99% of the expected return for the same risk. Does the solution of T10 seem to be more or less sensitive than the solution to T9?

2. Compare the performance of `P-SUMT`, `AL-SUMT` and `AL-SQP` on instances of problem (16.3.1) – (16.3.4) based on solutions to problems T12 and T14.

Barriers

You hit it, running.
The dry-stone wall deflects you
down its stubborn length

till the hillside tilt
releases you or else a
second obstacle,

lurking in a dip,
destroys more hope of getting
where you want to be.

Chapter 17

INEQUALITY CONSTRAINED OPTIMIZATION

1. Portfolio problems with inequality constraints

In chapter 9 we considered versions of the minimum-risk problem which exclude short-selling or place upper limits on invested fractions. So far, we have dealt with such problems by means of transformations of the variables like $y_i = x_i^2$. A better way of dealing with such restricions is to add *inequality constraints* in the form of bounds on the variables. The minimum-risk problem without short-selling can be written as

Minrisk2

$$\text{Minimize} \quad F(y) = y^T Q y \tag{17.1.1}$$

$$\text{subject to} \quad c_1(y) = \frac{1}{R_p}(\bar{r}^T y - R_p) = 0 \tag{17.1.2}$$

$$\text{and} \quad c_2(y) = e^T y - 1 = 0 \tag{17.1.3}$$

$$\text{and} \quad c_{i+2}(y) = y_i \geq 0 \quad \text{for } i = 1, ..., n \tag{17.1.4}$$

Minimum-risk problems involving more general linear inequality constraints can also arise if we wish to restrict a portfolio so as to avoid over-investment in one sector. Suppose, for instance, that the first k of the assets being considered are supermarket shares and that we do not want more than 25% of our portfolio to be in this area. To include this condition in a minimum-risk (which also excludes short-selling) we use the following problem.

$$\text{Minimize} \quad F(y) = y^T Q y \tag{17.1.5}$$

$$\text{subject to} \quad c_1(y) = \frac{1}{R_p}(\bar{r}^T y - R_p) = 0 \tag{17.1.6}$$

$$\text{and} \quad c_2(y) = e^T y - 1 = 0 \tag{17.1.7}$$

$$\text{and} \quad c_{2+i}(y) = y_i \geq 0 \quad \text{for} \quad i = 1,..,n \tag{17.1.8}$$

$$\text{and} \quad c_{n+3} = 0.25 - \sum_{i=1}^{k} y_i \geq 0 \tag{17.1.9}$$

Obviously there is a maximum-return problem similar to **Minrisk2**, namely

Maxret2

$$\text{Minimize} \quad F(y) = -\bar{r}^T y \tag{17.1.10}$$

$$\text{subject to} \quad c_1(y) = \frac{1}{V_a}(y^T Q y - V_a) = 0 \tag{17.1.11}$$

$$\text{and} \quad c_2(y) = e^T y - 1 = 0 \tag{17.1.12}$$

$$\text{and} \quad c_{i+2}(y) = y_i \geq 0 \quad \text{for} \quad i = 1,..,n \tag{17.1.13}$$

Sometimes the basic constraints on portfolio return or acceptable risk can usefully be expressed as inequalities. Consider the problem

$$\text{Minimize} \quad F(y) = y^T Q y \tag{17.1.14}$$

$$\text{subject to} \quad c_1(y) = \frac{1}{R_p}(\bar{r}^T y - R_p) \geq 0 \tag{17.1.15}$$

$$\text{and} \quad c_2(y) = e^T y - 1 = 0. \tag{17.1.16}$$

This minimizes risk subject to the portfolio return being *not less than* some target amount. It is a more flexible formulation than **Minrisk1** because (as we have seen) it sometimes happens that an increase in expected return corresponds to a decrease in risk. Similarly we could modify **Maxret1** so that portfolio return is maximized, subject to an *upper limit* on risk. This gives more flexibility than an approach which fixes an acceptable risk V_a.

$$\text{Minimize} \quad F(y) = -\bar{r}^T y \tag{17.1.17}$$

$$\text{subject to} \quad c_1(y) = \frac{1}{V_a}(V_a - y^T Q y) \geq 0 \qquad (17.1.18)$$

$$\text{and} \quad c_2(y) = e^T y - 1 = 0 \qquad (17.1.19)$$

Even the constraint $e^T y = 1$ can be relaxed to an inequality. In practice we do not have to invest all available capital and it might be advantageous to leave some of it as cash. Thus we could formulate the following minimum-risk problem which uses the modified risk, $\bar{V}$, defined by (16.1.2) and in which all the constraints are inequalities.

Minrisk4

$$\text{Minimize} \quad F(y) = \bar{V} = \frac{y^T Q y}{e^T y^2} \qquad (17.1.20)$$

$$\text{subject to} \quad c_1(y) = \frac{1}{R_p}(\bar{r}^T y - R_p) \geq 0 \qquad (17.1.21)$$

$$\text{and} \quad c_2(y) = 1 - e^T y \geq 0 \qquad (17.1.22)$$

$$\text{and} \quad c_{i+2}(y) = y_i \geq 0, \quad i = 1,..,n \qquad (17.1.23)$$

All the problems quoted so far are examples of the general constrained minimization or *nonlinear programming* problem. In general we write it as

$$\text{Minimize } F(x) \qquad (17.1.24)$$

$$\text{subject to} \quad c_i(x) = 0, \quad i = 1,..,l \qquad (17.1.25)$$

$$\text{and} \quad c_i(x) \geq 0, \quad i = l+1,..,m. \qquad (17.1.26)$$

Definition If F and all the c_i in (17.1.24) - (17.1.26) are linear functions then this is a *Linear Programming* (LP) problem. Specialised solution methods for this case can be found in [1].

Definition If, in (17.1.24) - (17.1.26), the function F is quadratic and the c_i are linear then it is a *Quadratic Programming* (QP) problem.

Definition If x satisfies the equality and inequality constraints (17.1.25),(17.1.26) it is said to be *feasible*. Otherwise it is called *infeasible*.

Exercise

Write the problem of maximizing return on a group of n assets subject to the constraint that risk must not exceed 1.5 times the risk level at the solution of **Minrisk0** and where short-selling in the first j assets is not allowed. How does the formulation change if we are prepared to leave some capital as cash?

2. Optimality conditions

Conditions which hold at a solution of (17.1.24)–(17.1.26) are extensions of those we have already considered for equality constrained problems.

Proposition If x^* is a local solution to (17.1.24) - (17.1.26) then the optimality conditions are as follows. The solution x^* must be feasible, and so

$$c_i(x^*) = 0, \quad i = 1,..,l \tag{17.2.1}$$

$$c_i(x^*) \geq 0, \quad i = l+1,..,m. \tag{17.2.2}$$

Furthermore, the Lagrange multipliers λ_i^*, $i = 1,...,m$ associated with the constraints must satisfy

$$\nabla L(x^*, \lambda^*) = \nabla F(x^*) - \sum_{i=1}^{m} \lambda_i^* \nabla c_i(x^*) = 0 \tag{17.2.3}$$

$$\lambda_i^* c_i(x^*) = 0 \quad (i = l+1,...,m) \tag{17.2.4}$$

$$\text{and} \quad \lambda_i^* \geq 0 \quad (i = l+1,...,m). \tag{17.2.5}$$

Definition Condition (17.2.4) is called the *complementarity* condition.
It states that an inequality constraint is *either* satisfied as an equality at x^* *or* it has a zero Lagrange multiplier.
Definition If $l+1 \leq i \leq m$ and $c_i(x^*) = 0$ we say that the i-th inequality is *binding*, and x^* lies on an *edge* of the feasible region.
Definition If $\lambda_i^* = 0$ when $l+1 \leq i \leq m$ then the i-th inequality is said to be *non-binding* and x^* is *inside* the i-th constraint boundary.

The non-negativity condition (17.2.5) on the Lagrange multipliers for the inequality constraints ensures that the function F will not be reduced by a move off any of the binding constraints at x^* to the interior of the feasible region.

The uniqueness of the Lagrange multipliers, λ^* depends on the normals to the *binding* constraints at x^* being linearly independent.

Second-order optimality conditions for (17.1.24)–(17.1.26) involve feasible directions for the binding constraints at x^*. Let I^* be the set of indices

$$I^* = \{i \mid l+1 \leq i \leq m \text{ and } c_i(x^*) = 0\} \tag{17.2.6}$$

and let N be the matrix whose first l rows are $\nabla c_1(x^*)^T, .., \nabla c_l(x^*)^T$ and whose remaining rows are $\nabla c_i(x^*)^T$ for $i \in I^*$. Then a second-order condition for x^* to be a solution of (17.1.24)–(17.1.26) is

$$z^T \nabla^2 L(x^*, \lambda^*) \geq 0 \quad \text{for any } z \text{ such that } Nz = 0. \qquad (17.2.7)$$

(This reduces to $z^T \nabla^2 F(x^*) z \geq 0$ if all the constraints are linear.)

Exercise
Consider the problem

$$\text{Minimize} \quad F(x) \quad \text{subject to} \quad c_1(x) \geq 0.$$

Suppose that x^* and λ_1^* satisfy the optimality condition (17.2.4) and that $c_1(x^*) = 0$ but that $\lambda_1^* < 0$. Show that there is a feasible point $\tilde{x} = x^* + \delta$ for which $F(\tilde{x}) < F(x^*)$. What does this imply about the optimality of x^*?

A worked example

Consider the problem

$$\text{Minimize } F(x) = x_1^2 + 3x_2^2 \qquad (17.2.8)$$

$$\text{subject to } c_1(x) = x_1 + 5x_2 - 1 \geq 0 \text{ and } c_2(x) = x_1 + 1 \geq 0. \qquad (17.2.9)$$

The solution values x_1^*, x_2^*, λ_1^*, λ_2^* must satisfy

$$x_1 + 5x_2 - 1 \geq 0 \text{ and } x_1 + 1 \geq 0 \qquad (17.2.10)$$

$$2x_1 - \lambda_1 - \lambda_2 = 0 \text{ and } 6x_2 - 5\lambda_1 = 0 \qquad (17.2.11)$$

$$\lambda_1(x_1 + 5x_2 - 1) = 0 \text{ and } \lambda_2(x_1 + 1) = 0 \qquad (17.2.12)$$

$$\lambda_1 \geq 0 \text{ and } \lambda_2 \geq 0 \qquad (17.2.13)$$

Rather than attempting to solve this system of equations and inequalities we shall simply use it to test the optimality of some candidate solutions.

Suppose first that we try the point $x_1 = -1$, $x_2 = \frac{2}{5}$ at which both constraints are binding. From (17.2.11), the corresponding Lagrange multipliers are

$$\lambda_1 = \frac{12}{25} \text{ and } \lambda_2 = -\frac{62}{25}.$$

This violates (17.2.13) so we deduce that $(-1,\ \frac{2}{5})$ is *not* a solution.

Next we investigate the possibility of a solution at $x_1 = -1,\ x_2 = 1$ with the second constraint binding but not the first. This will mean that $\lambda_1 = 0$. But if we put $\lambda_1 = 0$ in the first equation of (17.2.11) then it gives $\lambda_2 = 2x_1 = -2$ which violates the condition(17.2.13). Alternatively, if we consider the second equation in (17.2.11) it implies that $\lambda_1 = \frac{6}{5}$ which conflicts with the fact that λ_1 must be zero. Thus the optimality tests fail on *two* counts and so the feasible point $(-1,1)$ is not a solution.

Finally we consider whether there is a solution with the first constraint binding but not the second. This will mean that $\lambda_2 = 0$ and then (17.2.11) implies

$$2x_1 - \lambda_1 = 0 \text{ and } 6x_2 - 5\lambda_1 = 0.$$

For these equations to be consistent we need $2x_1 = \frac{6}{5}x_2$; and combining this with the first equation in (17.2.10) we get $x_1 = \frac{3}{28},\ x_2 = \frac{5}{28}$. It then follows that $\lambda_1 = \frac{6}{28},\ \lambda_2 = 0$ and all the first order optimality conditions are satisfied.

Exercise
By considering the first-order optimality conditions for the problem

$$\text{Minimize} \quad -3x_1 - 4x_2$$

$$\text{subject to} \quad x_1 \geq 0; \quad x_2 \geq 0; \quad 1 - x_1^2 - x_2^2 \geq 0$$

determine which (if any) of the following points is a solution:

(i) $x_1 = x_2 = 0$; (ii) $x_1 = 1,\ x_2 = 0$; (iii) $x_1 = \frac{4}{5},\ x_2 = \frac{3}{5}$

3. Transforming inequalities to equalities

A problem with inequality constraints can be transformed into one with only equalities if we introduce extra variables. Thus the problem

$$\text{Minimize} \quad F(x) \quad \text{subject to} \quad c_i(x) \geq 0, \quad i = 1,...,m \tag{17.3.1}$$

can be re-written as

$$\text{Minimize} \quad F(x) \quad \text{subject to} \quad c_i(x) - w_i^2 = 0, \quad i = 1,...,m. \tag{17.3.2}$$

Here $w_1,...,w_m$ are called *squared slack variables*. The conversion of (17.3.1) into (17.3.2) can have certain benefits when both n, the number of variables, and m, the number of constraints are quite small. However, for larger problems, the fact that (17.3.2) involves $n+m$ variables is usually a disadvantage. In the next chapter we introduce methods for handling inequalities directly.

Exercises

1. Solve the problem

$$\text{Minimize} \quad x_1^2 + x_2 \quad \text{subject to} \quad x_1 + x_2 \geq 3$$

by using a squared slack variable to transform it into an equality constrained problem. Show that the solution to the transformed problem satisfies the optimality conditions for the original one.

2. Show that, in general, the optimality conditions for (17.3.1) are consistent with those for (17.3.2).

3. Write down an optimization problem which uses squared slack variables to transform **Maxret2** into one involving just equality constraints. What is the relationship between the Lagrange multipliers for the two problems?

4. Transforming inequalities to simple bounds

The simplest forms of inequality constraint are simple bounds on the variables (as in **Minrisk2** and **Maxret2**). A problem with general inequality constraints can be transformed into one with only equality constraints and simple bounds. Thus the problem (17.3.1) can be re-written as

$$\text{Minimize} \quad F(x) \tag{17.4.1}$$

$$\text{subject to} \;\; c_i(x) - w_i = 0, \; i = 1, ..., m \;\; \text{and} \;\; w_i \geq 0, \; i = 1, ..., m. \tag{17.4.2}$$

In (17.4.2), $w_1, ..., w_m$ are called *slack variables*. The conversion of (17.3.1) into (17.4.1),(17.4.2) has the advantage that the inequalities are simple enough to be handled efficiently by the reduced gradient approach, as outlined in the next chapter. However, when m is large, (17.4.1), (17.4.2) involves many more variables than (17.3.1). In subsequent chapters, we shall consider methods which deal with (17.3.1) more directly.

Exercises

1. Solve the problem

$$\text{Minimize} \quad x_1^2 + x_2 \quad \text{subject to} \quad x_1 + x_2 > 3$$

using a slack variable to convert the inequality into an equation and a simple bound.

2. Show that the optimality conditions for (17.4.1), (17.4.2) are equivalent to those for (17.3.1).

5. Example problems

We now introduce some problems involving inequality constraints for use as examples in subseqent chapters. They are all based upon the problems T11-T14 from section 5 of chapter 12.

Problem T11a is the same as problem T11 but includes the lower bounds $y_i \geq 0$ to exclude short selling. The solution of problem T11a is

$$y \approx (0.333,\ 0,\ 0,\ 0,\ 0.667) \quad \text{with } V \approx 0.00667. \tag{17.5.1}$$

In contrast to solution of problem T11, we have all $y_i \geq 0$ but V is larger.

Problem T12a is the same as problem T12 with the addition of positivity constraints on the invested fractions. Its solution is

$$y \approx (0.245,\ 0.132,\ 0.242,\ 0,\ 0.031,\ 0.061,\ 0.037,\ 0,\ 0.184,\ 0.066) \tag{17.5.2}$$

giving $V \approx 0.0355$. This differs from the solution of problem T12 in having zero y_i in place of negative ones.

Problem T13a is the same as problem T13 but includes the lower bounds $y_i \geq 0$. Its solution is

$$y \approx (0.493,\ 0.255,\ 0,\ 0,\ 0.252) \quad \text{with } R \approx 1.103\%. \tag{17.5.3}$$

This expected return is slightly lower than that for problem T13.

Problem T14a is the same as problem T14 except for the inclusion of constraints $y_i \geq 0$. It has the solution

$$y \approx (0.19,\ 0.16,\ 0.3,\ 0,\ 0.062,\ 0.029,\ 0.005,\ 0.041,\ 0.167,\ 0.046) \tag{17.5.4}$$

which gives a smaller expected return than the soluton of problem T14.

Problem T11b is the same as problem T11 but includes the lower bounds $y_i \geq 0$ and also treats the expected-return and normalisation constraints as inequalities as in **Minrisk4**. This problem admits (at least) *two* solutions:

$$y \approx (0.476,\ 0.217,\ 0,\ 0,\ 0.306) \text{ with } \bar{V} \approx 0.00108 \text{ and } R \approx 1.09\% \tag{17.5.5}$$

$$y \approx (0.437,\ 0.2,\ 0,\ 0,\ 0.28)) \text{ with } \bar{V} \approx 0.00108 \text{ and } R \approx 1\% \tag{17.5.6}$$

Because the formulation **Minrisk4** replaces an equality constraint on expected return by a lower bound, the solution (17.5.5) gives a portfolio return which is *greater than* the target value $R_p = 1\%$. By doing this, it is also able to get a much smaller portfolio risk than is obtained at the solutions of T11 and T11a.

At the second solution (17.5.6), however, the expected return is precisely equal to the target value $R_p = 1\%$ but the sum of invested fractions is about 0.92 and so it is the constraint $e^T y \leq 1$ which is non-binding.

We would probably say that solution (17.5.5) is "better" than (17.5.6) since it gives a portfolio with a larger expected return. However, it is important to point out that *in terms of the problem as actually posed* both solutions are equally good. Both solutions are feasible points which return the same value of the objective function $\bar{V} = 0.00108$. The fact that we "prefer" the higher value of expected return is not actually reflected in the problem formulation and so both solutions are equally acceptable.

Problem T12b is derived from problem T12a by changing the normalisation and expected-return constraints to inequalities as in **Minrisk4**. This problem also has (at least) two valid solutions:

$$y \approx (0.2,\ 0.156, 0.284,\ 0,\ 0.055,\ 0.039,\ 0.012,\ 0.025,\ 0.17,\ 0.053) \quad (17.5.7)$$

with $\bar{V} \approx 0.02365$ and $r \approx 0.1772\%$; and

$$y \approx (0.115,\ 0.09,\ 0.16,\ 0,\ 0.031,\ 0.022,\ 0.007,\ 0.014,\ 0.095,\ 0.03) \quad (17.5.8)$$

with $\bar{V} \approx 0.02365,\ R = 0.1\%$.

Problem T13b is the same as problem T13 but includes the lower bounds $y_i \geq 0$ and treats all the constraints as inequalities as in
Maxret4

$$\text{Minimize} \quad F(y) = -\bar{r}^T y \quad (17.5.9)$$

$$\text{subject to} \quad c_1(y) = \frac{1}{V_a}(V_a - \frac{y^T Q y}{e^T y^2}) \geq 0 \quad (17.5.10)$$

$$\text{and} \quad c_2(y) = 1 - e^T y \geq 0 \quad (17.5.11)$$

$$\text{and} \quad c_{i+2}(y) = y_i \geq 0, \quad i = 1, ..., n. \quad (17.5.12)$$

This is the maximum-return problem corresponding to **Minrisk4**. The first two inequalities turn out to be binding and so Problem T13b has a unique solution which is the same as (17.5.3).

Problem T14b is also a maximum-return problem of the form **Maxret4**, using the same data as problem T14. Once again, the inequalities on risk and $\sum y_i$ turn out to be binding and the solution is given by (17.5.4).

Slack variables

Like those labourers
outside the vineyard, they need
opportunities.

They are positive:
whatever the vacancy
they're equal to it.

Chapter 18

EXTENDING EQUALITY-CONSTRAINT METHODS TO INEQUALITIES

1. Inequality constrained quadratic programming

When (17.1.24)–(17.1.26) is an inequality constrained quadratic programming problem (IQP), a solution procedure can be based on repeated use of the optimality conditions for an EQP. To simplify the explanation we describe the approach for an IQP with inequalities *only*; but it can be extended to deal with equalities as well. Consider the problem

$$\text{Minimize } \frac{1}{2}(x^T Gx) + h^T x + c \tag{18.1.1}$$

$$\text{subject to } \hat{A}x + \hat{b} \geq 0. \tag{18.1.2}$$

We shall assume first that G is positive definite.

Suppose we identify an *active set* as an estimate of those constraints which are binding at the solution. We might, for example, simply guess that the first t rows of $\hat{A}$ and $\hat{b}$ correspond to the active constraints. (We can make more informed choices than this!) If we now treat the active constraints as a set of equations *and ignore all other constraints* then we can obtain a trial solution $(\tilde{x}, \tilde{\lambda})$ by minimizing (18.1.1) subject to

$$Ax + b = 0.$$

We can do this by solving the EQP optimality conditions (13.1.4). If we find

$$\hat{A}\tilde{x} + \hat{b} \geq 0,$$

(so that $\tilde{x}$ does not violate any non-active constraints) and if

$$\tilde{\lambda}^T(\hat{A}\tilde{x} + \hat{b}) = 0 \text{ and } \tilde{\lambda} \geq 0$$

then optimality conditions (17.2.1)–(17.2.5) are all satisfied and the IQP is solved by $x^* = \tilde{x}$, $\lambda^* = \tilde{\lambda}$. If, however, $\tilde{x}$ and $\tilde{\lambda}$ are not optimal we must change the active set and solve another EQP. This process can be repeated until the active set becomes the binding set for problem (18.1.1) - (18.1.2).

The algorithm given below formalises the idea we have just outlined and shows how the active set can be systematically adjusted every time an EQP solution violates a new constraint or has a Lagrange multiplier whose sign indicates that an active constraint is non-binding. We use $\hat{a}_i$ to denote the i-th row of $\hat{A}$.

Inequality QP algorithm for a positive definite Hessian

Choose an initial point x and set $\lambda_1 = \ldots = \lambda_m = 0$.
Repeat
Identify the *active* constraints – i.e. those for which

$$\hat{a}_i^T x + \hat{b}_i < 0 \quad \text{or} \quad (\hat{a}_i^T x + \hat{b}_i = 0 \text{ and } \lambda_i \geq 0)$$

Renumber constraints so the active set is $i = 1,\ldots,t$
Set $g = Gx + h$ and $b_i = \hat{a}_i^T x + \hat{b}_i$ for $i = 1,\ldots,t$
Find p and μ to solve the EQP

$$\text{Minimize} \quad \frac{1}{2}(p^T G p) + g^T p \quad \text{s.t.} \quad \hat{a}_i^T p + b_i = 0 \quad i = 1,\ldots,t \tag{18.1.3}$$

Set $s = 1$, $\lambda_i = \mu_i$ $(i = 1,\ldots,t)$ and $\lambda_i = 0$ $(i = t+1,\ldots,m)$
Repeat for $i = t+1,\ldots,m$ (i.e. for all *inactive* constraints)

$$\text{if } \hat{a}_i^T p < 0 \quad \text{set } s = \min\left(s, -\frac{(\hat{a}_i^T x + \hat{b}_i)}{\hat{a}_i^T p}\right) \tag{18.1.4}$$

Replace x by $x + sp$
until optimality conditions are satisfied, i.e.

$$\hat{A}x + \hat{b} \geq 0 \quad Gx + h - \hat{A}^T \lambda = 0 \quad \lambda \geq 0$$

Implementations of this approach may differ in the method of solving (18.1.3) and also in rules about how many constraints may be dropped on a single iteration. The step size calculation (18.1.4) checks all the inactive constraints that might be violated by a step along p and ensures that no more than one constraint can be added to the active set on the current iteration. It may be advisable to allow only one constraint at a time to be deleted from the active set. (For a fuller discussion see Fletcher [54].)

We now consider the possibility that G in (18.1.1) may be indefinite. This complicates the active set approach because, even when the original problem

(18.1.1), (18.1.2) has a unique solution, it may happen that (18.1.3) cannot be solved for some choices of active set. One way to deal with this is to use the reduced gradient method for solving the EQP subproblem. If Z^TGZ in (13.3.6) is not positive definite then negative diagonal terms will appear during an attempt to calculate its Cholesky factors. As we explained in chapter 6, there are variants of the Cholesky method (e.g., [16]) which correct such negative terms and hence implicitly create a *modified* reduced Hessian which is then used to give a descent direction for the objective function. Further details of this are outside the scope of this section.

A worked example

Consider the IQP problem

$$\text{Minimize } F(x) = x_1^2 + 3x_2^2$$

$$\text{subject to } c_1(x) = x_1 + 5x_2 - 1 \geq 0 \text{ and } c_2(x) = x_2 \geq 0.$$

If we choose $x = (0,0)^T$ as the starting point then both constraints are active. Using the notation in the IQP algorithm, $g = (0,0)^T$ and $b = (-1,0)^T$ and we obtain a search direction p by solving the EQP subproblem

$$\text{Minimize } p_1^2 + 3p_2^2$$

$$\text{subject to } p_1 + 5p_2 - 1 = 0 \text{ and } p_2 = 0.$$

Here p is determined entirely by the constraints as $(1,0)^T$ and we get the Lagrange multipliers for the subproblem by solving

$$2p_1 - \lambda_1 = 0; \quad 6p_2 - 5\lambda_1 - \lambda_2 = 0$$

which gives

$$\lambda_1 = 2, \quad \lambda_2 = -5\lambda_1 = -10.$$

Hence the first iteration gives a new solution estimate

$$x_1 = 1, \ x_2 = 0, \ \lambda_1 = 2, \ \lambda_2 = -10.$$

which is feasible but non-optimal since λ_2 is negative.

For the next iteration we drop the second constraint from the active set. Now $g = \nabla F(x) = (2,0)^T$ and $b = 0$ and the EQP subproblem is

$$\text{Minimize } p_1^2 + 3p_2^2 + 2p_1$$

$$\text{subject to } p_1 + 5p_2 = 0.$$

Now p_1, p_2 and the Lagrange multiplier λ_1 must satisfy

$$p_1 + 5p_2 = 0; \ \ 2p_1 + 2 - \lambda_1 = 0; \ \ 6p_2 - 5\lambda_1 = 0.$$

Eliminating λ_1 between the second two equations we get

$$10p_1 - 6p_2 = -10$$

and so $p_1 = -\frac{25}{28}$, $p_2 = \frac{5}{28}$. We also get $\lambda_1 = \frac{6}{28}$.

Hence the result of the second iteration is the estimated solution

$$x_1 = \frac{3}{28}, \ x_2 = \frac{5}{28}, \ \lambda_1 = \frac{6}{28} \text{ and } \lambda_2 = 0.$$

It is left to the reader to show that this satisfies all the optimality conditions for the original problem.

Exercise
Use the IQP algorithm to solve the example problem above starting from
(i) the feasible initial guess $x = (1,1)^T$
(ii) the infeasible guess $x = (2.5, \ -0.2)^T$.

A minimum-risk example

Consider the minimum-risk problem problem (12.1.4)–(12.1.6) which is a QP with one equality and one inequality. Suppose we assume that both constraints are in the active set. Then we have already seen (chapter 12) that the solution of the corresponding EQP has Lagrange multipliers

$$\lambda_1 = \frac{\beta - \alpha}{\delta} \quad \lambda_2 = \frac{\beta - \gamma}{\delta}$$

where

$$\alpha = \frac{e^T Q^{-1} e}{2}, \quad \beta = \frac{\bar{r}^T Q^{-1} e}{2R_p}, \quad \gamma = \frac{\bar{r}^T Q^{-1} \bar{r}}{2R_p^2}, \quad \delta = \beta^2 - 2\alpha\gamma.$$

The corresponding invested fractions are given by

$$y^* = \frac{1}{2} Q^{-1} [\frac{\lambda_1}{R_p} \bar{r} - \lambda_2 e].$$

If it turns out that $\lambda_1 \leq 0$ then we can conclude that the inequality constraint is not binding at the solution. In this case, we can drop the first constraint from the active set and then (12.1.4)–(12.1.6) simply becomes **Minrisk0**.

Exercise
Solve the minimum-risk problem

$$\text{Minimize} \quad y^T Q y \quad \text{subject to} \quad \frac{\bar{r}^T y}{R_p} \geq 1 \quad \text{and} \quad e^T y = 1$$

where $\bar{r} = (1.2833,\ 1.1167,\ 1.0833)^T$,

$$Q = \begin{pmatrix} 0.0181 & -0.0281 & -0.00194 \\ -0.0281 & 0.0514 & 0.00528 \\ -0.00194 & 0.00528 & 0.0147 \end{pmatrix}$$

and $R_p = 1.0\%$. How does the solution change if $R_p = 1.2\%$?

2. Reduced gradients for inequality constraints

We can combine the reduced gradient approach with an active-set strategy in order to solve problems with a non-quadratic objective function and *linear* inequality constraints. However, because the active set keeps changing, it is necessary on each iteration to re-compute the Y and Z basis matrices matrices (see section 3 of chapter 13).

Reduced gradient algorithm for linear inequality constraints

Choose an initial *feasible* point x_0 and set $\lambda_0 = 0$
Choose B_0 as a positive definite estimate of $\nabla^2 F(x_0)$.
Set $g_0 = \nabla F(x_0)$
Repeat for $k = 0, 1, 2, ..$
 Determine active constraints – i.e., those with $c_i(x_k) = 0$ and $\lambda_{k_i} \geq 0$
 Get A_k as the matrix of active constraint normals at x_k
 Obtain Y_k and Z_k as basis matrices for the range and null spaces of A_k
 Determine z from $Z_k^T B_k Z_k z = -Z_k^T g_k$ and set $p_k = Z_k z$
 Find λ_{k+1} by solving $Y_k^T A_k^T \lambda = Y_k^T g_k + Y_k^T B_k p_k$
 Perform a line search to get $x_{k+1} = x_k + s p_k$. Find $g_{k+1} = \nabla F(x_{k+1})$.
 Do a quasi-Newton update of B_k with $\delta = x_{k+1} - x_k$, $\gamma = g_{k+1} - g_k$
until $||Z^T g_{k+1}||$ is less than a specified tolerance.

The line search step in this algorithm must be modified – as in the IQP algorithm above – so that the new point x_{k+1} does not violate any new constraints but allows at most one new constraint to become binding. Hence the line search will in general be a weak one in terms of the reduction in $F(x)$.

The algorithm given above can be extended to deal with nonlinear inequalities if a restoration step is included, as described in section 1 of chapter 14.

Reduced gradient methods for simple bounds

One case where Z is easy to calculate (which makes the reduced gradient approach very attractive) is when a constrained optimization problem involves no equalities and all the inequalities are *simple bounds*

$$l_i \leq x_i \leq u_i \quad i = 1,...,n.$$

In this situation we can split the variables at the start of each iteration into those which are "fixed" – i.e., on their bounds – and those which are "free". The Z matrix whose columns span the space of the free variables can then be taken simply as a partition of the identity matrix.

Reduced gradient algorithm for simple bounds

Choose an initial feasible point x_0
Choose B_0 as a positive definite estimate of $\nabla^2 F(x_0)$.
Set $g_0 = \nabla F(x_0)$
Repeat for $k = 0, 1, 2, ..$
Set Z_k to be the $n \times n$ identity matrix
Repeat for $i = 1, .., n$
If bounds on x_{k_i} are *active* bounds, i.e. if

$$(x_{k_i} = l_i \text{ and } g_{k_i} > 0) \quad \text{or} \quad (x_{k_i} = u_i \text{ and } g_{k_i} < 0)$$

then delete the i-th column of Z_k.
Set $p_k = Z_k z$ where $Z_k^T B_k Z_k z = -Z_k^T g_k$
Use a line search to find α so that $F(x_k + \alpha p_k) < F(x_k)$
Repeat for each free variable x_{k_i}

$$\sigma_i = \begin{cases} (u_i - x_{k_i})/p_{k_i} & \text{if } p_{k_i} > 0 \\ (l_i - x_{k_i})/p_{k_i} & \text{if } p_{k_i} < 0 \end{cases}$$

Set $x_{k+1} = x_k + s p_k$, where $s = \min(\alpha, \sigma_{t+1}, ..., \sigma_m)$.
Update B_k using $\delta = x_{k+1} - x_k$ and $\gamma = g_{k+1} - g_k$
until $||Z_k^T g_{k+1}|| <$ specified tolerance.

Exercise
Consider the problem

$$\text{Minimize} \quad F(y) = y^T Q y + 100(e^T y - 1)^2 \quad \text{subject to} \quad y_i \geq 0, \; i = 1, .., n$$

where $n = 3$ and

$$Q = \begin{pmatrix} 0.0181 & -0.0281 & -0.00194 \\ -0.0281 & 0.0514 & 0.00528 \\ -0.00194 & 0.00528 & 0.0147 \end{pmatrix}.$$

Use the reduced gradient approach to solve this problem, starting from the guess $y_1 = y_2 = 0$, $y_3 = 1$ and *taking B as the true Hessian* $\nabla^2 F$.

Numerical results with SOLVER

The version of the reduced-gradient method [41] implemented in SOLVER [7] can be applied to inequality constrained QPs and also to more general problems with inequality constraints. As mentioned before, SOLVER is more flexible than the reduced-gradient algorithm given above and it can be started at an infeasible point. The first few iterations locate a feasible point and then the algorithm proceeds as outlined in the previous section.

Table 18.1 shows the number of iterations needed by SOLVER to converge for variants of the linearly-constrained problems T11 and T12. (The bracketed figures show how many iterations are needed to give a feasible point.)

Starting guess	T11a	T11b	T12a	T12b
$y_i = \frac{1}{n}$	6(3)	8(0)	15(1)	16(1)
$y_i = 10^{-4}$	8(4)	11(1)	17(5)	26(5)

Table 18.1. Iteration counts for SOLVER on problems T11a–T12b

The increasing numbers of iterations show that the problems become more difficult as the number of inequality constraints gets larger. We note that SOLVER finds solution (17.5.5) for problem T11b when started from the standard guess $y_i = 0.2$. However, it converges to the second solution (17.5.6) when the initial point is $y_i = 0.0001$.

SOLVER can deal with nonlinear inequality constraints and so a similar set of performance figures for problems T13a–T14b appears in Table 18.2. The starting guess used is the standard one $y_i = \frac{1}{n}$.

T13a	T13b	T14a	T14b
18(3)	11(3)	48(8)	27(8)

Table 18.2. Iteration counts for SOLVER on problems T13a–T14b

The figures in Table 18.2 can be compared with those for other methods which are given later in this chapter. However, in view of the possibility of slow convergence of the reduced-gradient approach when applied to nonlinear equality constraints (section 1 in chapter 14), we shall focus our attention on the extension of SUMT and SQP methods to deal with nonlinear inequalities.

Exercises (To be solved using SOLVER or other suitable software.)
1. Let $y^{(a)}$ and $y^{(b)}$ denote the two solutions (17.5.5) and (17.5.6) for problem T11b. Show that any portfolio given by $y = cy^{(a)} + (1-c)y^{(b)}$ will give the same value for the risk function $\bar{V}$ given in (16.1.2).

2. Comment on the differences (if any) between the solutions of problems T12, T12a and T12b.

3. Show that problems T1, T1a and T1b all give the same solution – i.e., the addition of lower bounds on the y_i and the use of inequalities instead of equality constraints on portfolio return and $\sum y_i$ does not change the optimal invested fractions. Explain why problem T1 is different from T11, where the 'a' and 'b' variants *did* yield different solutions.

4. Solve problems T2a and T2b obtained by putting problem T2 in the forms **Minrisk2** and **Minrisk4.** Also solve a version of problem T1 in which the invested fractions are subject to upper and lower bounds $0.05 \leq y_i \leq 0.5$.

5. Add bounds $0 \leq y_i \leq 0.3$ on the invested fractions in the problems discussed in sections 1, 2 and 3 and solve them using the data for problems T11–T14.

3. Penalty functions for inequality constraints

For nonlinear inequality constraints, the P-SUMT approach can be applied, using a modified form of penalty function.
Definition A version of penalty function (14.2.1) for (17.1.24) - (17.1.26) is

$$P(x,r) = F(x) + \frac{1}{r}\{\sum_{i=1}^{l} c_i(x)^2 + \sum_{i=l+1}^{m} \min[0, c_i(x)]^2\}. \qquad (18.3.1)$$

The first penalty term treats the equalities as in (14.2.1) while the second includes contributions from the *violated* inequalities only. The function (18.3.1) can be used in an algorithm like P-SUMT because of the following result.
Proposition Suppose that (17.1.24)–(17.1.26) has a unique solution x^*, λ^* and that $F(x)$ is bounded below for all x. Suppose also that ρ is a positive constant and that, for all $r_k < \rho$ the Hessian matrix $\nabla^2 P(x, r_k)$ of (18.3.1) is positive definite for all x. If x_k denotes the solution of the unconstrained problem

$$\text{Minimize } P(x, r_k). \qquad (18.3.2)$$

then $x_k \to x^*$ as $r_k \to 0$. Furthermore, if $c_i(x_k) \leq 0$ as $r_k \to 0$,

$$\lambda_i^* = \lim_{r_k \to 0} \{-\frac{2}{r_k} c_i(x_k)\}. \qquad (18.3.3)$$

If $c_i(x_k) > 0$ as $r_k \to 0$ then $\lambda_i^* = 0$.

The essentials of the proof of (18.3.3) are similar to that for (14.2.3), (14.2.4) for the equality constraints. For more details and stronger results see [42].

Because (18.3.1) gathers all the constraints together in the penalty term, it does not explicitly identify an active set of constraints on every iteration.

A worked example

Consider the problem

$$\text{Minimize } x_1^2 + 2x_2^2 \text{ subject to } x_1 + x_2 \geq 2.$$

The penalty function for this problem is

$$P(x,r) = x_1^2 + 2x_2^2 + \frac{1}{r}[min(0, x_1 + x_2 - 2)]^2.$$

We need to consider whether, for $r > 0$, $P(x,r)$ can have a minimum at a feasible point. For this to happen we need

$$\frac{\partial P}{\partial x_1} = 2x_1 = 0 \text{ and } \frac{\partial P}{\partial x_2} = 4x_2 = 0$$

which implies $x_1 = x_2 = 0$ which is *not* a feasible point. Therefore the minimum of P will be at an infeasible point and

$$\frac{\partial P}{\partial x_1} = 2x_1 + \frac{2}{r}(x_1 + x_2 - 2) = 0 \text{ and } \frac{\partial P}{\partial x_2} = 4x_2 + \frac{2}{r}(x_1 + x_2 - 2) = 0.$$

This implies that $x_1 = 2x_2$ and therefore

$$4x_2 + \frac{2}{r}(3x_2 - 2) = 0.$$

and hence

$$x_2 = \frac{2}{2r+3}, \quad x_1 = \frac{4}{2r+3}.$$

In the limit, as $r \to 0$, the solution of the original problem is at $x = (\frac{4}{3}, \frac{2}{3})$.

Exercises

1. Write down expressions for the gradient and Hessian of the function (18.3.1). Explain why (18.3.1) may be harder to minimize than the penalty function (14.2.1) for equality constraints only – even when F is quadratic and all the c_i are linear.

2. Use the penalty function (18.3.1) to solve the problem

$$\text{Minimize } x_1^2 + x_2^2 - 2x_1 + 1 \quad \text{subject to } x_1 \geq 2.$$

How would the solution change if the constraint were $x_1 \geq 0$?

4. AL-SUMT for inequality constraints

The penalty function (18.3.1) can be difficult to minimize when r is very small for the reasons discussed in chapter 14. However the Augmented Lagrangian function can be extended to deal with inequality constraints and leads to a SUMT approach which does not require the penalty parameter to tend to zero.

Definition The Augmented Lagrangian function $M(x, v, r)$ for use with inequality constraints has the following form [55]

$$F(x)+\frac{1}{r}\{\sum_{i=1}^{l}(c_i(x)-\frac{r}{2}v_i)^2+\sum_{i=l+1}^{m}[min(0,c_i(x)-\frac{r}{2}v_i)]^2\}. \qquad (18.4.1)$$

This function has a stationary point at x^* and it can be used in the AL-SUMT approach. There are only two alterations to the algorithm in section 3 of chapter 14. The initial choices of the *inequality-constraint* parameters $v_{l+1},..,v_m$ must be non-negative and the rule for updating them is

$$v_{k+1,i}=\begin{cases} v_{k,i}-\frac{2}{r_k}c_i(x_k) & \text{if } c_i(x_k)<\frac{r_k}{2}v_{k,i} \\ 0 & \text{otherwise} \end{cases} \qquad (18.4.2)$$

,for $i=l+1,..,m$. Formula (18.4.2) will cause $v_k \to \lambda^*$ as $c(x_k) \to 0$. The justification for these changes is fairly straightforward [55].

Exercises

1. Apply AL-SUMT to the problem

$$\text{Minimize } x_1^2+2x_2^2 \text{ subject to } x_1+x_2 \geq 2.$$

using $v=1$ as the initial guess for the multiplier parameter.

2. There is an exact penalty function for the inequality constrained problem similar to the one given in chapter 14 for equality constrained problems. It is

$$E(x,r)=F(x)+\frac{1}{r}\{\sum_{i=1}^{l}|c_i(x)|+\sum_{i=l+1}^{m}|min(0,c_i(x))|\}. \qquad (18.4.3)$$

Use this function to solve the problem in question 1.

5. SQP for inequality constraints

The ideas in chapter 15 can be extended to deal with (17.1.24) - (17.1.26) simply by including the inequality constraints (in linearised form) in the QP subproblem. Thus, in WHP-SQP, the section which gives the search direction

and new trial Lagrange multipliers is
Obtain p_k and λ_{k+1} by solving the QP subproblem

$$\text{Minimize } \frac{1}{2}p^T B_k p + \nabla F(x_k)^T p$$

$$\text{subject to } c_i(x_k) + \nabla c_i(x_k)^T p = 0 \quad i = 1,...,l$$

$$c_i(x_k) + \nabla c_i(x_k)^T p \geq 0 \quad i = l+1,...,m.$$

The line search along p_k will involve a function such as (18.4.1) or (18.4.3) which handles inequality constraints.

Similarly, in AL-SQP, p_k and u_k are obtained by solving the QP subproblem

$$\text{Minimize } \frac{1}{2}p^T H_k^{-1} p + p^T \nabla F(x)$$

$$\text{subject to } c_i(x_k) + \nabla c_i(x_k)^T p = -\frac{r_k}{2}(u_{k_i} - \lambda_{k_i}) \quad i = 1,...,l$$

$$\text{and } c_i(x_k) + \nabla c_i(x_k)^T p \geq -\frac{r_k}{2}(u_{k_i} - \lambda_{k_i}) \quad i = l+1,...,m$$

where H_k is an estimate of the inverse Hessian of the Lagrangian. This subproblem approximates the Newton direction towards the minimum of the Augmented Lagrangian (18.4.1) (see [49]). The line search in the inequality-constraint version of AL-SQP is also based on obtaining a reduction in (18.4.1).

The IQP subproblems in both algorithms can be solved by an active-set approach as outlined in section 1 of this chapter. This strategy will be simplified if we ensure that the updated Hessian approximation matrix remains positive definite throughout. When solving the EQPs which occur inside the IQP subproblems it can be more efficient to work with an updated estimate of the *inverse* Hessian of the Lagrangian as in (15.3.13) and (15.3.14).

6. Results with `P-SUMT`, `AL-SUMT` and `AL-SQP`

The `SAMPO` program `sample10` can apply SUMT and SQP methods to **Minrisk2** and **Maxret2** in which the inequality constraints are simple lower bounds on the invested fractions to prevent short-selling. Program `sample11` applies the same constrained techniques to **Minrisk4** and **Maxret4**. By using these programs we obtain Tables 18.3 and 18.4 which show the numbers of iterations and function values needed by SUMT and SQP methods on a set of test problems. For comparison, we also quote the number of `SOLVER` iterations (but not function calls) from Tables 18.1 and 18.2. The standard starting guess $y_i = 1/n$ is used in each case.

Method	T11a	12a	T13a	T14a
P-SUMT	84/324	85/207	152/373	160/378
AL-SUMT	65/233	48/106	68/107	128/235
AL-SQP	10/15	12/18	11/15	39/49
SOLVER	6/–	15/–	18/–	48/–

Table 18.3. Perfromace of SUMT and SQP on problems T11a–T14a

Method	T11b	T12b	T13b	T14b
P-SUMT	25/114	126/526	135/353	277/606
AL-SUMT	24/112	87/341	59/136	166/322
AL-SQP	8/18	15/22	11/15	40/97
SOLVER	8/–	16/–	11/–	27/–

Table 18.4. Performace of SUMT and SQP on problems T11b–T14b

In these tables, AL-SQP and SOLVER show comparable performance. AL-SQP also seems much better than either of the SUMT methods. However, when we are dealing with inequality constrained problems, this last remark must be interpreted with some caution. Each iteration of AL-SQP involves an IQP solution which may in turn involve many EQP steps. For instance, if an iteration starts from a feasible point and if the IQP subproblem has t binding constraints then at least t EQP problems will be solved, each one involving the solution of a system of equations like (13.1.4). In practice, the early iterations of AL-SQP are usually more expensive than the later ones. Once the iterates are fairly close to x^* the IQP subproblem will have a "warm start" so that the initial active set is the same (or nearly the same) as the binding set and only one or two EQP steps will be needed. However, on average, each iteration of AL-SQP *may* involve a good deal more linear algebra than an iteration of P-SUMT or AL-SUMT. Each iteration of a SUMT method merely involves the algebra associated with computing a search direction and updating an inverse-Hessian approximation: and, in this respect, there is no difference between equality- and inequality-constrained problems.

Exercises (To be solved using `sample10`, `sample11` or other suitable software. In particular, the P-SUMT and AL-SUMT approaches can be implemented by applying SOLVER to minimize $P(x,r)$ or $M(x,v,r)$ for a sequence of values of the parameters r and v.)

1. Construct a table similar to 18.4 which compares SUMT and SQP approaches on problems T1a–T6a.

2. Do all the methods give the same solution for problems T11b and T12b? Discuss.

3. The figures in Table 18.4 were obtained using weak line searches in the unconstrained minimizations: construct a similar table for the case when `P-SUMT` and `AL-SUMT` use `QNp` rather than `QNw`.

4. If, for any data set, the solution to **Minrisk0** does not involve short-selling and gives an expected return R_{min} show that the solution to (17.1.20) - (17.1.23) will be the same as the solution to **Minrisk0** when $R_p < R_{min}$.

5. Suppose that, for some dataset, **Minrisk2** and **Minrisk4** yield the same invested fractions. Explain why the calcuated values of Lagrange multipliers may be different.

6. Re-write **Minrisk2** and **Maxret2** as nonlinear equality-constrained problems using squared slack variables. Hence solve problems T11a and T13a via this formulation and discuss the results.

7. Can the `SUMT` and `SQP` approaches find *both* solutions to problems T11b and T12b? (Investigate this by using starting guesses very close to both solutions given in section 5 of chapter 17.)

Complementarity

Face it. One must go.
This town just ain't big enough
for the both of us.

Chapter 19

BARRIER FUNCTION METHODS

1. Introduction

Penalty function methods generate a sequence of *infeasible* points $\{x_k\}$ which come closer to the constraints as the iterations proceed. By contrast, barrier function methods – *which can only be applied to problems with inequalities but no equality constraints* – generate points which lie inside the feasible region. Hence, for the remainder of the chapter, we consider the problem

$$\text{Minimize } F(x) \tag{19.1.1}$$

$$\text{subject to } \; c_i(x) \geq 0 \;\; (i = 1, ..., m) \tag{19.1.2}$$

2. Barrier functions

Definition One form of barrier function for the problem (19.1.1), (19.1.2) is

$$B(x, r) = F(x) + r \sum_{i=1}^{m} \frac{1}{c_i(x)}. \tag{19.2.1}$$

Because the barrier term includes reciprocals of the constraints we see that B is very much greater than F when any $c_i(x)$ is near zero – i.e. when x is near the boundary of the feasible region. Similarly, $B \approx F$ when all the $c_i(x)$ are much greater than zero and x is in the interior of the feasible region.

Definition A second (more popular) barrier function for (19.1.1), (19.1.2) is

$$B(x,\ r) = F(x) - r\sum_{i=1}^{m} \log(c_i(x)). \tag{19.2.2}$$

When $1 > c_i(x) > 0$ then $\log(c_i(x) < 0$. Hence the second term on the right of (19.2.2) implies $B >> F$ when any of the constraint functions approaches zero. Note, however, that (19.2.2) is undefined when any $c_i(x)$ is negative.

There is a relationship, similar to that for penalty functions, between unconstrained minima of $B(x,r)$ and the solution if (19.1.1), (19.1.2).

Proposition Suppose that (19.1.1), (19.1.2) has a unique solution $x,^*, \lambda^*$. Suppose also that ρ is a positive constant and, for all $r_k < \rho$, the Hessian matrix $\nabla^2 B(x,r_k)$ of the barrier functions (19.2.1) or (19.2.2) is positive definite for all x. If x_k denotes the solution of the unconstrained problem

$$\text{Minimize } B(x,r_k) \tag{19.2.3}$$

then

$$x_k \to x^* \text{ as } r_k \to 0.$$

Moreover, if B is defined by (19.2.1),

$$\frac{r_k}{c_i(x_k)^2} \to \lambda_i^* \text{ as } r_k \to 0; \tag{19.2.4}$$

and if B is defined by (19.2.2),

$$\frac{r_k}{c_i(x_k)} \to \lambda_i^* \text{ as } r_k \to 0. \tag{19.2.5}$$

We omit the main part of the proof of this result. However it is easy to justify (19.2.4) because differentiating (19.2.1) gives

$$\nabla B(x_k,r_k) = \nabla F(x_k) - \sum_{i=1}^{m} \frac{r_k}{c_i(x_k)^2}\nabla c_i(x_k) = 0. \tag{19.2.6}$$

By comparing (19.2.6) with the Lagrangian stationarity condition (12.2.2) as $r_k \to 0$ we deduce (19.2.4). A similar argument justifies (19.2.5).

This proposition is the basis of the B-SUMT algorithm, stated below. (A fuller theoretical background can be found in [42].) The convergence test for the algorithm is based on satisfying the complementarity condition (17.2.4), using the estimated Lagrange multipliers implied by (19.2.4) or (19.2.5). B-SUMT can often be used successfully, in practice, for problems of the form (19.1.1), (19.1.2) even when the conditions in the proposition cannot be verified.

Barrier Function SUMT (B-SUMT)

Choose an initial guessed solution x_0
Choose a penalty parameter r_1 and a constant $\beta(< 1)$
Repeat for $k = 1,2,...$
starting from x_{k-1} use an iterative method to find x_k to solve (19.2.3)
set $r_{k+1} = \beta r_k$
if B is defined by (19.2.1) then

$$\hat{\lambda}_i = \frac{r_k}{c_i(x_k)^2} \quad \text{for } i = 1,..,m$$

else, if B is defined by (19.2.2) then

$$\hat{\lambda}_i = \frac{r_k}{c_i(x_k)} \quad \text{for } i = 1,..,m$$

until $\hat{\lambda}_1 c_1(x_k),..,\hat{\lambda}_m c_m(x_k)$ are all sufficiently small.

Exercise
Obtain expressions for the Hessian matrices of barrier functions (19.2.1), (19.2.2). Hence find an expression for the Newton search direction for (19.2.2). How could this expression be modified if $(\nabla^2 F(x))^{-1}$ were available?

A worked example

As an example of the use of the log-barrier function we consider the problem

$$\text{Minimize } F(x) = x_1^2 + 3x_2^2 \tag{19.2.7}$$

$$\text{subject to } \; c_1(x) = x_1 + 5x_2 - 1 \geq 0. \tag{19.2.8}$$

The corresponding barrier function is

$$B(x,r) = x_1^2 + 3x_2^2 - r\log(x_1 + 5x_2 - 1)$$

and hence the minimum of $B(x,r)$ satisfies

$$\frac{\partial B}{\partial x_1} = 2x_1 - \frac{r}{(x_1 + 5x_2 - 1)} = 0 \tag{19.2.9}$$

$$\frac{\partial B}{\partial x_2} = 6x_2 - \frac{5r}{(x_1 + 5x_2 - 1)} = 0. \tag{19.2.10}$$

Eliminating the term involving r between these two equations we get

$$x_2 = \frac{5}{3}x_1; \tag{19.2.11}$$

and substitution in (19.2.9) gives

$$2x_1(x_1 + \frac{25}{3}x_1 - 1) - r = 0.$$

This simplifies to

$$\frac{56}{3}x_1^2 - 2x_1 - r = 0 \qquad (19.2.12)$$

so that

$$x_1 = \frac{3}{112}(2 \pm \sqrt{(4 + \frac{224r}{3}})).$$

Using (19.2.11) we get

$$x_2 = \frac{5}{112}(2 \pm \sqrt{(4 + \frac{224r}{3}})).$$

In these expressions the quantity under the square root is greater than 4 when $r > 0$. Hence (19.2.12) gives one positive and one negative value for x_1. But (19.2.11) means that x_2 must have the same sign as x_1. However, a solution with *both* x_1 and x_2 negative cannot satisfy (19.2.8). Therefore the unconstrained minimum of $B(x,r)$ is at

$$x_1 = \frac{3}{112}(2 + \sqrt{(4 + \frac{224r}{3}})), \quad x_2 = \frac{5}{112}(2 + \sqrt{(4 + \frac{224r}{3}})).$$

Hence, as $r \to 0$ we have $x_1 \to \frac{3}{28}$ and $x_2 \to \frac{5}{28}$. These values satisfy the optimality conditions for problem (19.2.7), (19.2.8).

Exercises

1. Deduce the Lagrange multiplier for the worked example above.

2. Use a log-barrier function approach to solve the problem

$$\text{Minimize} \quad x_1 + x_2 \quad \text{subject to} \quad x_1^2 + x_2^2 \le 2$$

3. A log-barrier approach is used to solve the problem

$$\text{Minimize} \quad -\bar{r}^T y \quad \text{subject to } y^T Q y \le V_a.$$

Suppose that the barrier parameter r is chosen so the minimum of $B(y,r)$ occurs where $y^T Q y = kV_a$, where $k < 1$. Obtain an expression for $y(r)$ which minimizes the barrier function and hence find r in terms of $\bar{r}$, Q, and V_a.

4. Solve

$$\text{Minimize} \quad x_1 + 2x_2 \quad \text{subject to} \quad x_1 \ge 0,\ x_2 \ge 1$$

using the barrier function (19.2.1).

3. Numerical results with B-SUMT

We use B-SUMT to denote the SAMPO implementation of the barrier SUMT algorithm using the log-barrier function (19.2.2). In B-SUMT the unconstrained minimizations are done by QNw or QNp.

A safeguard is needed in the line search for the unconstrained minimization technique used in B-SUMT. The log-barrier function is undefined if any of the constraints $c_i(x)$ are non-positive and therefore the line search must reject trial points where this occurs. This can be done within the framework of the Armijo line-search by re-setting $B(x,r)$ to a very large value at any point x which violates one or more of the constraints.

In this section we consider the minimum-risk and maximum-return problems in the forms **Minrisk4** and **Maxret4**. Program sample11 allows us to solve such problems using B-SUMT and we begin with Problem T11b. We note first that B-SUMT *must be started with a feasible point.* Hence the automatic initial guess $y_i = 1/n,\ i = 1,...,n$ (suitable for the other two SUMT algorithms) will probably not be appropriate for B-SUMT. In the present case, it is fairly easy to find a feasible starting guess. Since we have the expected return for each asset (4.4.4), we can set $y_3 = 0.95$ and $y_i = 0.01,\ i = 1,2,4,5$. This ensures $\sum y_i < 1$ and also that the expected portfolio return exceeds $R_p(1\%)$. The progress made by B-SUMT is shown in Table 19.1. For comparison, this table also summarises the behaviour of P-SUMT from the same starting point. In both cases the unconstrained minimizer was QNw and the initial penalty parameter and rate of reduction were given by $r_0 = 0.1,\ \beta = 0.25$.

	B-SUMT			P-SUMT		
k	$F(x_k)$	r_k	QNw cost	$F(x_k)$	$\|\|c(x_k)\|\|$	QNw cost
1	2.09×10^{-3}	1.0×10^{-1}	31/47	1.08×10^{-3}	5.2×10^{-5}	30/63
2	2.07×10^{-3}	2.5×10^{-2}	37/56	1.08×10^{-3}	1.2×10^{-5}	33/76
3	1.99×10^{-3}	6.3×10^{-3}	48/69	1.08×10^{-3}	8.4×10^{-7}	34/84
4	1.75×10^{-3}	1.6×10^{-3}	63/87			
5	1.41×10^{-3}	3.9×10^{-4}	80/107			
6	1.20×10^{-3}	9.8×10^{-5}	99/128			
7	1.12×10^{-3}	2.4×10^{-5}	116/148			
8	1.09×10^{-3}	6.1×10^{-6}	131/167			
9	1.09×10^{-3}	1.5×10^{-6}	150/238			
10	1.08×10^{-3}	3.8×10^{-7}	158/273			

Table 19.1. B-SUMT and P-SUMT solutions to problem T11b

The fact that P-SUMT converges much more quickly than B-SUMT is largely due to the fact that the minimum of $B(x,0.1)$ is very much further from the optimum than the minimum of $P(x,0.1)$. For this problem, therefore, the initial choice $r_1 = 0.1$ is a bad one for the barrier approach.

The solution to problem T11b obtained with B-SUMT is approximately

$$y_1 \approx 0.46,\ y_2 \approx 0.21,\ y_3 = y_4 = 0,\ y_5 \approx 0.29$$

with $\bar{V} \approx 1.083$ and $R \approx 1.044\%$. The solution from P-SUMT is

$$y_1 \approx 0.45,\ y_2 \approx 0.2,\ y_3 = y_4 = 0,\ y_5 \approx 0.29$$

giving $\bar{V} \approx 1.083$ and $R \approx 1.026\%$. These both lie between the two solutions (17.5.5) and (17.5.6). Both solutions are both equally valid because they are feasible points giving the same value of the objective function $\bar{V}$. The fact that B-SUMT and P-SUMT terminate at different points seems to be due to the fact that one works inside the feasible region while the other operates outside.

We now apply B-SUMT to the maximum-return problem T13b. We need to find a feasible starting values for the y_i, which is not so straightforward as it was for **Minrisk4**. However, if we have already solved **Minrisk4** then its solution will be appropriate as an initial guess for **Maxret4** provided it gives $\bar{V} < V_a$. Alternatively, if we first solve **Minrisk0** then we must get a solution with $V < V_a$ (unless V_a has been chosen less than V_{min} so that *no* portfolio with the given data yields an acceptable risk.)

Since Problem T13b is based on the same asset data as Problem 11b we can use the feasible solution from B-SUMT as a starting point. Table 19.2 compares the rates of convergence of B-SUMT and P-SUMT.

	B-SUMT			P-SUMT		
k	$F(x_k)$	r_k	QNw cost	$F(x_k)$	$\|\|c(x_k)\|\|$	QNw cost
1	−1.02	1.0×10^{-1}	25/37	−1.16	5.5×10^{-2}	17/48
2	−1.07	2.5×10^{-2}	38/58	−1.12	1.4×10^{-2}	35/85
3	−1.09	6.3×10^{-3}	50/109	−1.11	3.4×10^{-3}	53/142
4	−1.10	1.6×10^{-3}	63/167	−1.10	8.6×10^{-4}	65/171
5	−1.10	3.9×10^{-4}	73/216	−1.10	2.2×10^{-4}	76/200
6	−1.10	9.8×10^{-5}	90/331	−1.10	5.4×10^{-5}	91/252
7	−1.10	2.4×10^{-5}	103/408	−1.10	1.3×10^{-5}	100/288
8	−1.10	6.1×10^{-6}	118/465	−1.10	3.4×10^{-6}	113/330

Table 19.2. B-SUMT and P-SUMT solutions to problem T13b

In this case the behaviour of the barrier approach is comparable with that of the penalty method. The initial unconstrained minima of $B(x,0.1)$ and $P(x,0.1)$

are at about the same distance from the true solution and hence B-SUMT and P-SUMT need about the same number of QNw iterations to converge. B-SUMT needs more function calls, however, because of the extra restrictions on step length mentioned at the beginning of this section. Since problem T13b has a unique solution, both methods converge to the same result.

Table 19.3 gives a comparison between the SUMT and SQP methods on problems T11b – T14b.

Method	T11b	T12b	T13b	T14b
P-SUMT	34/84	106/283	109/318	173/428
AL-SUMT	33/74	69/143	40/102	90/181
B-SUMT	183/445	155/382	113/451	157/513
AL-SQP	20/25	14/22	8/18	47/97

Table 19.3. Perfromance of SUMT and SQP on problems T11b–T14b

Because B-SUMT is included, the counts of iterations and function values are based on a (different) feasible starting guess for each problem and so the entries for P-SUMT, AL-SUMT and AL-SQP differ from those in Table 18.4. Specifically, these starting guesses are as follows:
For problem T11b: $y_i = 0.01$ for $i = 1,2,4,5$; $y_3 = 0.95$
For problem T12b: $y_i = 0.1$ for $i = 2,3,5,8,9,10$; $y_i = 0.001$ for $i = 1,4,6,7$
For problem T13b: y is given by the feasible solution to problem T11b.
For problem T14b: y is given by the feasible solution to problem T12b.

We can see that the barrier function approach is usually the least competitive of the SUMT methods. Hence, in the form described in this chapter, its use is normally confined to those problems where the function cannot be calculated at some infeasible points. A simple example would be if the expression for $F(x)$ included terms involving $\sqrt{x}_1, .., \sqrt{x}_n$ which are non-computable if the any of the constraints $x_1 \geq 0, .., x_n \geq 0$ are violated. In such situations it is important to use a method whose iterates stay inside the constraint boundaries.

In spite of the relatively poor performance of B-SUMT, the ideas behind the method are important because they are the foundation for the *interior point* methods described in the next chapter.

Exercises (To be solved using sample11 or other suitable software. In particular, B-SUMT can be implemented by applying SOLVER to minimize $B(x,r)$ for a sequence of values of r.)
1. Apply B-SUMT to problem T11b using other feasible starting guesses such as $y_1 = 0.99, y_i = 0.0001$, $i = 2,...,5$. Compare performance with that of

AL-SUMT from the same initial guess. Can you find a starting guess from which B-SUMT converges to the same solution as AL-SUMT?

2. By performing numerical experiments using problems T11b and T13b, investigate how the speed of convergence of B-SUMT varies with the choice of initial parameter r_1.

3. Use problems T2b and T4b to compare the performance of the log-barrier and reciprocal-barrier forms of the B-SUMT algorithm.

4. Consider the Lagrange multiplier estimates provided by B-SUMT, P-SUMT and AL-SUMT at the solutions to T11b and T13b and comment on any differences you observe.

Chapter 20

INTERIOR POINT METHODS

1. Introduction

Interior point methods are based on barrier functions. They are widely used for nonlinear programming, following their initial (and continuing) popularity for linear programming applications [56]. Consider the problem

$$\text{Minimize } F(x) \text{ subject to } c_i(x) \geq 0, \quad i = 1, \ldots, m \qquad (20.1.1)$$

We can use *slack variables* to reformulate the inequalities as equalities, giving

$$\text{Minimize } F(x) \text{ subject to } c_i(x) - w_i = 0, \quad i = 1, \ldots, m \qquad (20.1.2)$$

$$\text{where} \quad w_i \geq 0, \quad i = 1, \ldots, m \qquad (20.1.3)$$

Now suppose that we deal with the bounds on the w_i by a barrier term and consider the transformed problem, involving a parameter, $r (> 0)$

B-NLP

$$\text{Minimize } F(x) - r \sum_{i=1}^{m} \log(w_i) \qquad (20.1.4)$$

$$\text{subject to } c_i(x) - w_i = 0, \quad i = 1, \ldots, m. \qquad (20.1.5)$$

We now state an important property of problem B-NLP (which depends on quite mild conditions on F and the c_i).

Proposition If, for all r less than some constant ρ, problem (20.1.4), (20.1.5) has a unique solution with $\hat{x}(r)$, $\hat{w}(r)$ and $\hat{\lambda}(r)$ as the optimal values of the variables, slack variables and Lagrange multipliers respectively then, as $r \to 0$,

$$\{\hat{x}(r), \; \hat{w}(r), \; \hat{\lambda}(r)\} \to \{x^*, \; w^*, \; \lambda^*\},$$

where $\{x^*, w^*, \lambda^*\}$ solves (20.1.2).

A sequential *constrained* minimization technique would solve (20.1.4) for a decreasing sequence of r-values, and hence deduce the solution to (20.1.2). However – as in AL-SQP – we would like to avoid the cost of complete minimizations by simply approximating solutions of (20.1.4) in a way that causes them to become more accurate as r approaches zero.

Exercise
Form the problem B-NLP using data from T4b and solve it for a decreasing sequence of values of the parameter r.

2. Approximate solutions of problem B-NLP

The Lagrangian function associated with (20.1.4), (20.1.5) is

$$L(x,w,\lambda) = F(x) - r\sum_{i=1}^{m}\log(w_i) - \sum_{i=1}^{m}\lambda_i(c_i(x) - w_i). \qquad (20.2.1)$$

The first order optimality conditions at the solution $(\hat{x}, \hat{w}, \hat{\lambda})$ are:

$$c_i(\hat{x}) - \hat{w}_i = 0 \qquad i = 1,..,m; \qquad (20.2.2)$$

$$\nabla_x L = \nabla F(\hat{x}) - \sum_{i=1}^{m}\hat{\lambda}_i \nabla c_i(\hat{x}) = 0; \qquad (20.2.3)$$

$$\frac{\partial L}{\partial w_i} = -\frac{r}{\hat{w}_i} + \hat{\lambda}_i = 0 \qquad i = 1,..,m. \qquad (20.2.4)$$

Equation (20.2.2) ensures feasibility while (20.2.3) and (20.2.4) are stationarity conditions for the original variables and the slacks. In what follows, ∇ and ∇^2 operators without subscripts will always relate to differentiation with respect to the original x variables only.

Now suppose $(x,\ w,\ \lambda)$ is an approximate solution of (20.1.4) and we want to find δx, δw, $\delta\lambda$ so that $\hat{x} = x + \delta x$, $\hat{w} = w + \delta w$ and $\hat{\lambda} = \lambda + \delta\lambda$. From (20.2.2)

$$c_i(x + \delta x) - w_i - \delta w_i = 0 \qquad i = 1,..,m$$

and a first order Taylor approximation to $c_i(x + \delta x)$ gives

$$\nabla c_i(x)\delta x - \delta w_i = w_i - c_i(x) \quad i = 1,..,m. \qquad (20.2.5)$$

From (20.2.3)

$$\nabla F(x + \delta x) - \sum_{i=1}^{m}(\lambda_i + \delta\lambda_i)\nabla c_i(x + \delta x) = 0$$

and by using first order Taylor approximations of the gradient terms we get

$$\nabla F(x) + \nabla^2 F(x)\delta x - \sum_{i=1}^{m}(\lambda_i + \delta\lambda_i)(\nabla c_i(x) + \nabla^2 c_i(x)\delta x) = 0.$$

We ignore the second-order terms involving $\delta\lambda_i\delta x$ and combine the terms involving the Hessians of the function and constraints so that

$$\tilde{G} = \nabla^2 F(x) - \sum_{i=1}^{m} \lambda_i \nabla^2 c_i(x).$$

Thus we obtain

$$\nabla F(x) + \tilde{G}\delta x - \sum_{i=1}^{m}(\lambda_i + \delta\lambda_i)\nabla c_i(x) = 0$$

which rearranges as

$$\tilde{G}\delta x - \sum_{i=1}^{m} \delta\lambda_i \nabla c_i(x) = \sum_{i=1}^{m} \lambda_i \nabla c_i(x) - \nabla F(x). \qquad (20.2.6)$$

Finally, from (20.2.4)

$$(w_i + \delta w_i)(\lambda_i + \delta\lambda_i) = r \qquad i = 1,..,m.$$

Dropping the second-order term $\delta w_i \delta\lambda_i$ and re-arranging we obtain

$$\delta w_i = \frac{r}{\lambda_i} - w_i - w_i\frac{\delta\lambda_i}{\lambda_i} \qquad i = 1,..,m. \qquad (20.2.7)$$

Substituting for δw_i in (20.2.5) yields

$$\nabla c_i(x)\delta x + w_i\frac{\delta\lambda_i}{\lambda_i} = -c_i(x) + \frac{r}{\lambda_i} \qquad i = 1,..,m \qquad (20.2.8)$$

We now write $g = \nabla F(x)$ and let A denote the Jacobian matrix whose rows are $\nabla c_i(x)$, $i = 1,..,m$. We write W, Λ for the diagonal matrices whose elements are w_i and λ_i respectively. As usual, e denotes the vector with elements $e_i = 1$, $i = 1,..,m$. Then we can express (20.2.6) and (20.2.8) as a symmetric system of equations for δx and $\delta\lambda$.

$$\tilde{G}\delta x - A^T\delta\lambda = -g + A^T\lambda \qquad (20.2.9)$$

$$-A\delta x - W\Lambda^{-1}\delta\lambda = c - r\Lambda^{-1}e \qquad (20.2.10)$$

These equations have some similarity with (15.3.9) and (15.3.10) which give δx and $\delta\lambda$ in augmented Lagrangian SQP.

Once δx and $\delta\lambda$ have been found by solving (20.2.9), (20.2.10) we can recover δw from a re-arrangement of (20.2.7)

$$\delta w = r\Lambda^{-1}e - w - W\Lambda^{-1}\delta\lambda. \tag{20.2.11}$$

In a later section we shall describe a practical algorithm based on (20.2.9)-(20.2.11) in which (δx, δw) is regarded as a search direction along which an acceptable step must be determined.

A worked example

We consider the problem

$$\text{Minimize } x_1^2 + 3x_2^2 \text{ subject to } x_1 + 5x - 2 - 1 \geq 0,\ 5x_1 - x_2 - 0.25 \geq 0.$$

We start from $x = (0.25,\ 0.2)^T$ where $g = (0.5,\ 1.2)^T$ and $c = (0.25, 0.8)^T$. The Hessian and Jacobian matrices for the problem are

$$\tilde{G} = \begin{pmatrix} 2 & 0 \\ 0 & 6 \end{pmatrix} \text{ and } A = \begin{pmatrix} 1 & 5 \\ 5 & -1 \end{pmatrix}.$$

We can take $\lambda = (0.23, 0.054)^T$ as a convenient starting guess because this gives $g \approx A^T\lambda$. A suitable choice for w can then be based on the observation that, at a solution of B-NLP, $r = \lambda_i w_i$, for $i = 1,..,m$. Therefore, if we set $r = 0.005$ we can take $w = (0.0217,\ 0.0926)^T$. Since

$$W = \begin{pmatrix} 0.0217 & 0 \\ 0 & 0.0926 \end{pmatrix} \text{ and } \Lambda = \begin{pmatrix} 0.23 & 0 \\ 0 & 0.054 \end{pmatrix}$$

the equations (20.2.9), (20.2.10) for δx and $\delta\lambda$ now turn out to be

$$\begin{pmatrix} 2 & 0 & -1 & -5 \\ 0 & 6 & -5 & 1 \\ -1 & -5 & -0.0945 & 0 \\ -5 & 1 & 0 & -1.7147 \end{pmatrix} \begin{pmatrix} \delta x_1 \\ \delta x_2 \\ \delta\lambda_1 \\ \delta\lambda_2 \end{pmatrix} = \begin{pmatrix} 0 \\ -0.1040 \\ 0.2283 \\ 0.7074 \end{pmatrix}.$$

This gives

$$\delta x_1 \approx -0.129,\ \delta x_2 \approx -0.0197,\ \delta\lambda_1 \approx -0.0126,\ \delta\lambda_2 \approx -0.0489$$

and from (20.2.11) we get $\delta w_1 \approx 0.0012$, $\delta w_2 \approx 0.0839$. Hence the new point is $x \approx (0.1214,\ 0.1803)^T$ and the revised multipliers are $\lambda \approx (0.2174,\ 0.0051)^T$. Finally, the corrected slack variables are $w \approx (0.0229,\ 0.1765)^T$.

The solution of the original problem is at $x^* \approx (0.1071,\ 0.1786)^T$ with the first constraint binding but not the second. Hence the iteration has moved the variables appreciably closer to x^* and has also moved λ_2 closer to zero.

Exercises

1. Use (20.2.7) to eliminate $\delta\lambda$ instead of δw and obtain equations similar to (20.2.9), (20.2.10) with δx and δw as unknowns. By performing a suitable change of variable show that this can be made into a symmetric system.

2. If the inverse $\tilde{G}^{-1}$ is available, show that values of δx and $\delta\lambda$ which solve (20.2.9), (20.2.10) can be obtained from

$$\delta\lambda = (A\tilde{G}^{-1}A^T + W\Lambda^{-1})^{-1}(A\tilde{G}^{-1}g + r\Lambda^{-1}e - c - A\tilde{G}A^T\lambda)$$

$$\delta x = \tilde{G}^{-1}(-g + A^T(\lambda + \delta\lambda))$$

3. Do a second iteration of the worked example given above. What would have happened on the first iteration if we had chosen $r = 0.0025$?

3. An interior point algorithm

We can now give a simple outline of an interior point algorithm. The new values obtained on iteration k will be of the form

$$x_{k+1} = x_k + s\delta x_k, \quad w_{k+1} = w_k + s\delta w_k, \quad \lambda_{k+1} = \lambda_k + \delta\lambda_k$$

where δx_k, $\delta\lambda_k$ and δw_k are obtained by solving (20.2.9), (20.2.10) with all the coefficients and right-hand-side values evaluated at $(x_k,\ w_k,\ \lambda_k)$. The steplength s must be small enough for $w_k + s\delta w_k$ to be positive because we are only interested in feasible points of subproblem (20.1.4).

We also need to choose s so that $(x_k + s\delta x_k, w_k + s\delta w_k)$ is a better solution estimate than (x_k, w_k). We could, for example, perform a line search in terms of an Augmented Lagrangian for problem (20.1.4), namely

$$\hat{M}(x,w,v,r) = F(x) - r\sum_{i=1}^{m}\log(w_i) - (c(x) - w)^T v + \frac{||c(x) - w)||_2^2}{r} \qquad (20.3.1)$$

where v is a vector of Lagrange multiplier approximations. It can be shown that under certain circumstances the choice $v = \lambda_{k+1}$ will ensure that $(\delta x_k,\ \delta w_k)$ is a descent direction for (20.3.1) at (x_k, w_k), i.e. that

$$\delta x_k^T \nabla_x \hat{M}(x_k,\ w_k,\ v,\ r) + \delta w_k^T \nabla_w \hat{M}(x_k,\ w_k,\ v,\ r) < 0 \qquad (20.3.2)$$

However, it is not clear that, when x_k is far from x^*, the choice $v = \lambda_{k+1}$ will be a good enough estimate of the true multipliers for a line search with respect to (20.3.1) to be helpful for overall convergence. The algorithm IPM, given below, a strategy like that in AL-SQP which fixes "sensible" values for v and retains them for several iterations until they can safely be replaced after some tests on the errors in optimality conditions. When $v \neq \lambda_{k+1}$ the descent property (20.3.2) may be ensured if r is chosen sufficiently small on each iteration.

Algorithm IPM uses a quasi-Newton approach to update a matrix B_k to approximate $\tilde{G}$ (which is, in turn, an estimate of the Hessian of the Lagrangian function). Revision of λ and r is done as in AL-SQP using an error function based on the optimality conditions (20.2.2) - (20.2.4), namely

$$\tau(x,w,\lambda,r) = ||c(x)-w||^2 + ||g - A^T\lambda||^2 + ||W\Lambda e - re||^2 \qquad (20.3.3)$$

When τ is sufficiently small we can consider that we are sufficiently close to a solution of problem (20.1.4) to permit the penalty parameter r to be reduced.

Interior Point Algorithm (IPM)

Choose initial values x_0, $w_0 (> 0)$, $\lambda_0 (> 0)$, B_0, r_0 and v_0
Choose a scaling factor $\beta < 1$ and set $\tau_r^- = \tau(x_0, w_0, \lambda_0, r_0)$
Repeat for $k = 0,1,2,...$
Obtain δx_k and $\delta\lambda_k$ by solving

$$B_k\delta x - A_k^T\delta\lambda = -g_k + A_k^T\lambda_k$$

$$-A_k\delta x - W_k\Lambda_k^{-1}\delta\lambda = c_k - r_k\Lambda_k^{-1}e$$

Set $\delta w_k = r_k\Lambda_k^{-1}e - w_k - W_k\Lambda_k^{-1}\delta\lambda_k$, and $\bar{s} = 1$
Repeat for $i = 1,...,m$

$$\text{If } \delta w_{k_i} < 0 \text{ set } \bar{s} = \min(\bar{s}, -0.9\frac{w_{k_i}}{\delta w_{k_i}})$$

$$\text{Set } \lambda_i^+ = \max(10^{-6}, \lambda_{k_i} + \delta\lambda_{k_i})$$

Get $x_{k+1} = x_k + s\delta x_k$, $w_{k+1} = w_k + s\delta w_k$ ($s \leq \bar{s}$), by a line search to give

$$\hat{M}(x_{k+1}, w_{k+1}, v_k, r_k) < \hat{M}(x_k, w_k, v_k, r_k)$$

Obtain B_{k+1} by a quasi-Newton update of B_k
Set $\lambda_{k+1} = \lambda_k$, $r_{k+1} = r_k$ and $v_{k+1} = v_k$
If $\tau(x_{k+1}, w_{k+1}, \lambda^+, r_k) < \tau_r^-$ then
set $\tau_r^- = \tau(x_{k+1}, w_{k+1}, \lambda^+, r_k)$, $\lambda_{k+1} = \lambda^+$, $r_{k+1} = \beta r_k$ and $v_{k+1} = \lambda_{k+1}$
until $||\tau(x_{k+1}, w_{k+1}, \lambda_{k+1}, 0)||$ is sufficiently small Algorithm IPM implements the main feature of an interior point method – namely the use of a search direction on every iteration which is based on approximating the minimum of problem **B-NLP**. Many variations of the central idea have been suggested which get δx, δw and $\delta\lambda$ from equations other than (20.2.9)–(20.2.11) and which perform a line searches using merit functions other than (20.3.1). Accounts of some of these alternative algorithms can be found in [13] and [57], for instance.

Exercises

1. Suppose δx, δw, $\delta\lambda$ are obtained by solving (20.2.9)–(20.2.11). In order for $(\delta x, \delta w)$ to be a descent direction for the augmented Lagrangian $\hat{M}$ given by (20.3.1) we require

$$\delta x^T \nabla_x \hat{M} + \delta w^T \nabla_w \hat{M} < 0$$

where $\nabla_x \hat{M} = g + 2A^T\{2(c-w)/r - v\}$ and $\nabla_w \hat{M} = -rW^{-1}e + v - \frac{2}{r}(c-w)$.

Show that $\delta x^T \nabla_x \hat{M} + \delta w^T \nabla_w \hat{M}$ is equivalent to the expression

$$-\delta x^T \tilde{G} \delta x - \delta w^T \Lambda W^{-1} \delta w - \frac{2}{r}(c-w)^T(c-w) - (c-w)^T(\lambda^+ - v) \quad (20.3.4)$$

where $\lambda^+ = \lambda + \delta\lambda$.

2. Using (20.3.4), show that, if $\tilde{G}$ is positive definite and if $W^{-1}\Lambda$ is positive semi-definite then $(\delta x, \delta w)$ is a descent direction with respect to (20.3.1) for *any* value of r if $v = \lambda^+$. Show also that if $v \neq \lambda^+$ the descent property with respect to $\hat{M}$ may be ensured if r is chosen sufficiently small on each iteration.

3. Explain why (20.2.4) implies that, when the parameter r_k is replaced by βr_k, a good way to adjust the values of the slack variables w and the multipliers λ might be to use one of the following formulae for each $i = 1, \ldots, m$.

$$\text{if } \lambda_{k_i} < w_{k_i} \text{ then } \lambda_{(k+1)_i} = \beta\lambda_{k_i} \quad w_{(k+1)_i} = w_{k_i} \quad (20.3.5)$$

$$\text{else } w_{(k+1)_i} = \beta w_{k_i} \quad \lambda_{(k+1)_i} = \lambda_{k_i}. \quad (20.3.6)$$

4. Numerical results with IPM

The SAMPO implementation of the algorithm described in the previous section is called `IPM`. Program `sample11` allows us to apply `IPM` to minimum-risk and maximum-return problems. Table 20.1 shows progress of `IPM` iterations on problem T13b when $r_1 = 0.1$ and $\beta = 0.25$. Comparison with Table 19.2 shows that `IPM` converges more quickly than both `B-SUMT` and `P-SUMT`. This happens because `IPM` avoids explicit minimizations of the subproblems for each value of barrier parameter r.

Table 20.2 compares `IPM` with `AL-SQP` on the test examples T11b - T13b. Clearly `IPM` is much more competitive than any of the SUMT approaches and can outperform the SQP method on occasions.

k	$F(x_k)$	r_k	itns/function calls
1	-1.08	1.0×10^{-1}	4/9
2	-1.06	2.5×10^{-2}	7/12
3	-1.08	6.25×10^{-6}	10/15
4	-1.10	1.56×10^{-3}	18/29
5	-1.10	3.91×10^{-4}	25/40
6	-1.10	9.77×10^{-5}	32/52
7	-1.10	2.44×10^{-5}	39/63
8	-1.10	6.10×10^{-6}	45/70

Table 20.1. `IPM` solution to problem T13b

Method	T11b	T12b	T13b	T14b
`IPM`	33/45	40/61	42/65	33/53
`AL-SQP`	20/25	14/22	8/18	47/97

Table 20.2. Performance of `IPM` and `AL-SQP` on problems T11b–T14b

At the time of writing, interior point methods are the subject of much research and development and many algorithms have been proposed. Some are designed for special situations such as LP or QP problems. Those intended for the general (possibly non-convex) nonlinear programming problem include [58] and [59]. The success of such interior point methods has brought them into competition with implementations of the SQP approach and it still seems an open question which of these two approaches is "better". The computational cost of an IPM iteration can be similar to that for AL-SQP, since both methods get a search direction by solving a linear system obtained by approximating the optimality conditions for a perturbed form of the original minimization problem. The IPM system will include *all* inequality constraints and hence will usually be larger than the system used by AL-SQP which only involves constraints in the current active set. On the other hand, AL-SQP may have to solve several systems on each iteration until the correct active set is established. Gould et al [60]) suggest IPM and SQP can *co-exist* because IP algorithms can be an efficient way to solve the QP subproblems in SQP methods.

Exercise (To be solved using `sample11` or other suitable software.)
Problems T1b–T4b use the data for problems T1–T4 but include all inequality constraints in the manner of **Minrisk4** and **Maxret4**. Compare `IPM` with `AL-SQP` and `SOLVER` on the solution of thse problems.

Chapter 21

DATA FITTING USING INEQUALITY CONSTRAINTS

1. Minimax approximation

In chapter 11 we observed that one way of fitting a model $z = \phi(x,t)$ to a set of observations (t_i, z_i), $i = 1,...,m$ is to solve the *minimax* problem

$$\text{Minimize} \quad max_{1 \leq i \leq m} |z_i - \phi(x, t_i)|. \tag{21.1.1}$$

If (21.1.1) is viewed as an unconstrained optimization problem it is clearly non-smooth and therefore more difficult to deal with than the sum-of-squares functions arising in the least-squares approach (also described in chapter 11). However, we can also calculate minimax approximations using a differentiable constrained minimization problem. If there are n parameters x_i appearing in the model function ϕ then a solution to (21.1.1) can be found by solving

$$\text{Minimize} \quad x_{n+1} \tag{21.1.2}$$

$$\text{subject to} \quad x_{n+1} \geq z_i - \phi(x, t_i) \geq -x_{n+1} \quad i = 1,...,m. \tag{21.1.3}$$

As a simple illustration, we fit a model $z = \phi(x,t) = x_1 t$ to the data

$$t_1 = 1,\ z_1 = 1; \quad t_2 = 2,\ z_2 = 1.5.$$

The minimax solution is obtained from

$$\text{Minimize} \quad x_2$$

$$\text{subject to} \quad x_2 \geq 1 - x_1 \geq -x_2; \quad x_2 \geq 1.5 - 2x_1 \geq -x_2.$$

From a rough sketch it is clear that the solution line must pass *below* the first point and *above* the second. Therefore we expect the two binding constraints to be

$$x_2 - 1 + x_1 \geq 0 \quad \text{and} \quad 1.5 - 2x_1 + x_2 \geq 0.$$

Treating these inequalities as equations, we get $x_1 = 0.8333$ and $x_2 = 0.1667$. The two non-zero Lagrange multipliers λ_1, λ_2 must satisfy

$$-\lambda_1 + 2\lambda_2 = 0 \quad \text{and} \quad 1 - \lambda_1 - \lambda_2 = 0.$$

Hence $\lambda_1 = \frac{2}{3}$, $\lambda_2 = \frac{1}{3}$. The fact that these are both positive confirms that we have found the solution.

As a less trivial example, we fit a minimax straight line to the ten-point dataset $(t_i,\ z_i)$ (11.4.1) by solving problem (21.1.2), (21.1.3) with $n = 3$ and

$$\phi(x,\ t_i) = x_2 + x_1 t_i.$$

The `SAMPO` program `sample12` solves this problem using `AL-SQP` or `IPM` for the constrained minimizations. We obtain a feasible starting guess by finding z_{max} and z_{min} as the largest and smallest of the given z_i values and setting

$$x_1 = 0;\ x_2 = \frac{z_{max} + z_{min}}{2};\ x_3 = 1.01\frac{z_{max} - z_{min}}{2}.$$

From this starting point `AL-SQP` (with $r_1 = 0.1$, $\beta = 0.25$) solves (21.1.2), (21.1.3) in 6 iterations and 9 function calls, giving the minimax line as

$$z = 3.94 + 0.035t$$

which is similar to the least-squares line obtained in section 4 of chapter 11. `IPM` gets the same solution in 20 iterations (using 22 function calls).

Exercises

1. Find the straight line $z = a + bt$ that gives the minimax fit to the following points $(t_i,\ z_i)$:

$$(0, 0.1),\ (0.5,\ 0.5),\ (1, 0.9).$$

2. Solve the problem (21.1.2), (21.1.3) when ϕ is $z = x_3 + x_2 t + x_1 t^2$ and the data is given by (11.4.1).

2. Trend channels for time series data

Financial data, such as share-price histories, is typically "jagged" – i.e., its graph has many peaks and troughs, as shown in figure 21.1 which illustrates a typical 65-day history of closing prices of a certain share.

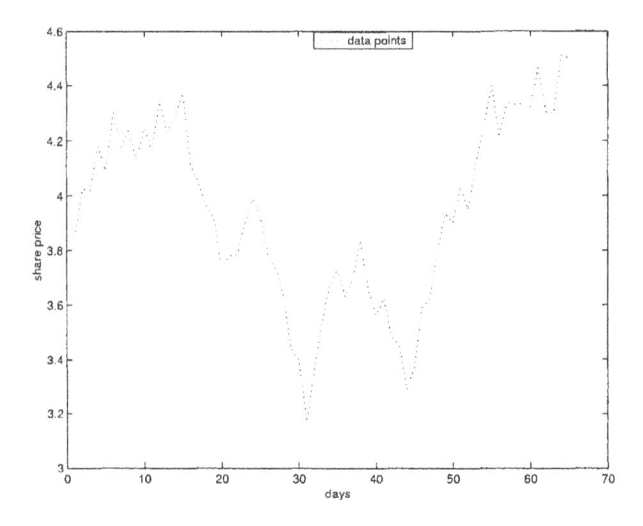

Figure 21.1. Typical share price data for 65 days

We can use least-squares or minimax approximation to calculate a *trend line* by fitting a model $z = a + bt$ to data $(t_i,\ z_i)$, $i = 1, ..., m$. This can be thought of as estimating an underlying "average" behaviour. The minimax trend line for the data in Figure 21.1 is $z = 3.7 + 0.00286t$. This is significantly different from least-squares trend line for the same data which is $z = 3.9 + 0.00105t$.

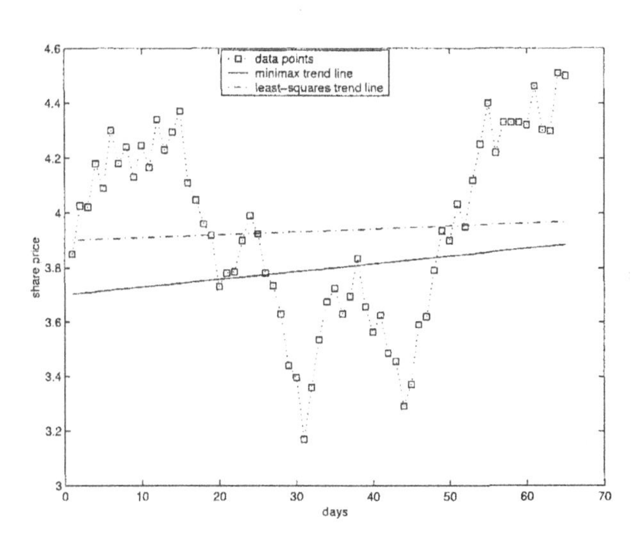

Figure 21.2. Trend lines for share price data in figure 21.1

Figure 21.2 shows that, relative to the least-squares line, the minimax approximation has been "pulled down" towards the two lowest data points. It is to be expected that minimax curve-fitting will be more senstive to extreme data values because it aims to reduce the *worst* discrepancies between the model and the data. The least-squares approach, on the other hand, seeks to minimize an overall measure of the errors.

Reducing a set of data points like those in figure 21.1 to a single trend line may be an over-simplification. Instead it may be useful to determine a *trend channel*. This aims to measure the spread of the data and involves finding a *pair* of parallel lines to enclose some or all of the given points as tightly as possible. The lower line may be called the *support line* while the upper one is known as the *resistance line*. If we write these, respectively, as

$$z = x_1 t + x_2 \quad \text{and} \quad z = x_1 t + x_3$$

then the problem we want to solve is

$$\text{Minimize} \quad (x_3 - x_2)^2 \tag{21.2.1}$$

$$\text{subject to} \quad x_1 t_i + x_3 \geq z_i \geq x_1 t_i + x_2 \quad i = 1, \ldots, m. \tag{21.2.2}$$

A feasible starting point for this problem is

$$x_1 = 0; \quad x_2 = z_{min}; \quad x_3 = z_{max}.$$

We now use this initial guess in problem (21.2.1), (21.2.2) to calculate a trend channel based on the first twenty points in the data shown in figure 21.1. The resulting support and resistance lines are

$$z = 3.86 - 0.00632t \quad \text{and} \quad z = 4.46 - 0.00632t$$

(This solution is obtained by `AL-SQP` in 5 iterations, using program `sample12`.)

Figure 21.3 shows the 20-day trend channel, projected forward to day 30. We observe that the actual share prices stay within the channel for seven days but then they "break out" and fall below the support line. This can be taken as an indication that it may be time to start selling holdings in this particular share. Of course, one would have to make a judgement on the severity of such a break out before choosing to take action. Consideration of the share-price history (Figure 21.1) between days 28 and 50 suggests that a decision to sell could well have been justified.

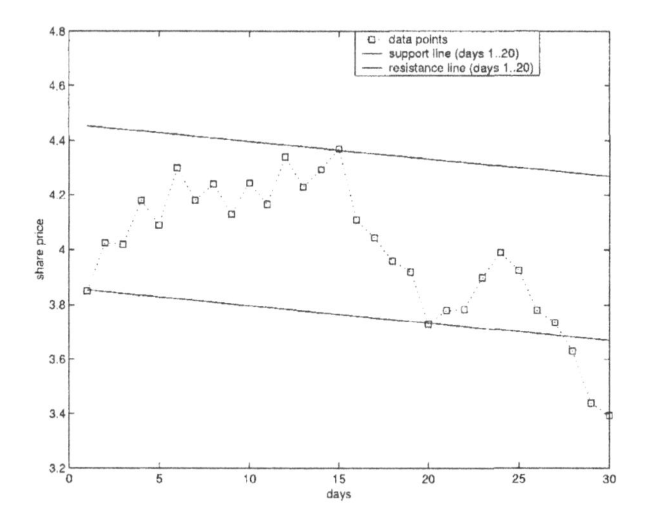

Figure 21.3. Trend channel (days 1 to 20) for data in Figure 21.1

Figure 21.4 shows a similar trend channel calculated for the share-price data between days 20 and 40.

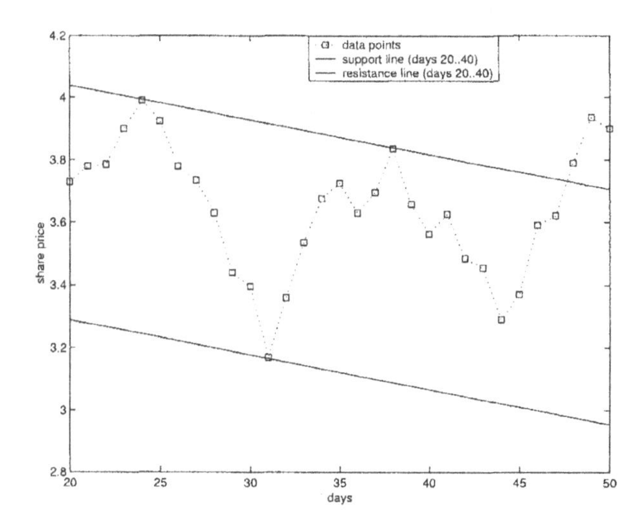

Figure 21.4. Trend channel (days 20 to 40) for data in Figure 21.1

In Figure 21.4 the trend channel is downwards and the share prices stay between the support and resistance lines until day 48. When breakout does occur, it is through the resistance line which suggests that shares should be purchased.

Figure 21.5 shows the channel based on the points for days 40 to 60.

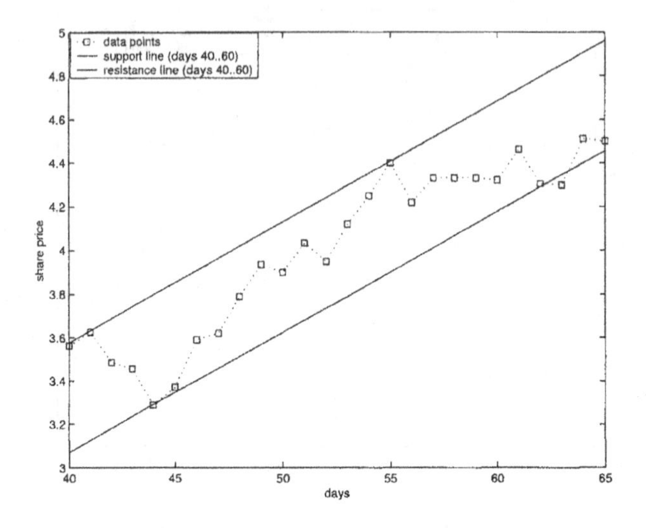

Figure 21.5. Trend channel (days 40 to 60) for data in Figure 21.1

In Figure 21.5 we have an upward trend channel and the available data up to day 65 indicates that we should maintain current holdings.

Exercises

1. Use `SOLVER` to calculate a trend channel for the ten-day share-price data (11.4.1). Given that the actual prices for days 11 – 15 are

$$4.165,\ 4.34,\ 4.23,\ 4.295,\ 4.37$$

what buying or selling action is indicated?

2. Explain why a suitable starting guess for problem (21.2.1), (21.2.2) can be obtained by the following procedure: *(i)* calculate the minimax trend line $z = a + bt$ and then *(ii)* set

$$x_1 = b; \quad x_2 = z_{min} - bt_{min}; \quad x_3 = z_{max} - bt_{max}$$

where (t_{min}, z_{min}), (t_{max}, z_{max}) are, respectively, the lowest and highest points in the dataset. By using `SOLVER` or `sample12` determine how well this starting point approximates the solutions of the trend channel problems illustrated in Figures 21.3 – 21.5.

Chapter 22

PORTFOLIO RE-BALANCING AND OTHER PROBLEMS

1. Re-balancing allowing for transaction costs

We now return to the treatment of transaction costs given in [52] which we have already considered in chapter 16.

Suppose we have an existing portfolio involving n assets and defined by invested fractions $\hat{y}_1,..,\hat{y}_n$. Suppose also that, over the most recent m time periods, the mean returns are given by the n-vector $\bar{r}$ and and that the variance-covariance matrix is Q. Re-balancing the portfolio to reflect the latest information may involve buying and/or selling some of the assets. We let x_i denote the fraction of the current portfolio worth by which $\hat{y}_i$ changes due to buying more of asset i. Similarly, x_{n+i} is the fraction of current portfolio worth by which $\hat{y}_i$ changes due to selling some holdings in asset i. We shall constrain these to be non-negative and so

$$x_i \geq 0 \text{ and } x_{n+i} \geq 0 \text{ for } i = 1,...,n.$$

The new invested fraction in asset i is $y_i = \hat{y}_i + x_i - x_{n+i}$. In what follows we shall exclude short-selling and so we require

$$y_i \geq 0, \quad i = 1,..,n.$$

If c_b is the unit transaction cost associated with buying any asset and c_s the unit transaction cost of making a sale then the fraction of the current portfolio worth that will be spent in transaction costs is

$$C_T = c_b \sum_{i=1}^{n} x_i + c_s \sum_{i=1}^{n} x_{n+i}. \tag{22.1.1}$$

Hence the sum of the new invested fractions will correspond to only a part of the current portfolio worth – that is

$$\sum_{i=1}^{n} y_i = (1 - C_T) \sum_{i=1}^{n} \hat{y}_i. \quad (22.1.2)$$

If $\sum_{i=1}^{n} \hat{y}_i = 1$ then clearly $\sum_{i=1}^{n} y_i < 1$ and therefore – as in chapter 16 – the risk associated with the new portfolio is given by

$$\bar{V} = \frac{y^T Q y}{(e^T y)^2}. \quad (22.1.3)$$

If we are re-balancing the portfolio in order to obtain minimum-risk for some target return then we require

$$\bar{r}^T y = R_p.$$

We can combine all the above into a nonlinear programming problem in terms of the variables $x_1, .., x_{2n}$. It will be convenient to write x^+ for $(x_1, .., x_n)^T$ and x^- for $(x_{n+1}, .., x_{2n})^T$. The current invested fractions $\hat{y} = (\hat{y}_1, .., \hat{y}_n)^T$ are given and we assume $e^T \hat{y} = 1$. Then from (22.1.1) and (22.1.2)

$$e^T y = (1 - C_T) = 1 - c_b e^T x^+ - c_s e^T x^-. \quad (22.1.4)$$

But $e^T y = e^T(\hat{y} + x^+ - x^-)$ and $e^T \hat{y} = 1$. Therefore we deduce

$$(1 + c_b) e^T x^+ - (1 - c_s) e^T x^- = 0. \quad (22.1.5)$$

Using (22.1.4) in the definition of risk (22.1.3), the objective function in the minimum-risk rebalancing problem is

$$F(x) = \bar{V} = \frac{(\hat{y} + x^+ - x^-)^T Q (\hat{y} + x^+ - x^-)}{(1 - c_b e^T x^+ - c_s e^T x^-)^2}. \quad (22.1.6)$$

There are two equality constraints. These are (22.1.5) and the target requirement for expected portfolio return

$$\bar{r}^T \hat{y} + \bar{r}^T (x^+ - x^-) - R_p = 0 \quad (22.1.7)$$

There are also $3n$ inequality constraints:
No short-selling

$$\hat{y} + x^+ - x^- \geq 0 \quad (22.1.8)$$

Positivity of variables

$$x^+ \geq 0; \quad x^- \geq 0. \quad (22.1.9)$$

All the constraints (22.1.5) and (22.1.7)–(22.1.9) are linear and so it is easy to obtain the constraint normals. We can write the gradient of the objective function (22.1.6) as $\nabla F = (g^+,\ g^-)^T$ to denote differentiation with respect to the buying and selling variables. A convenient way to obtain ∇F involves first regarding F as a function of y. From (22.1.3), the gradient with respect to y is

$$\bar{V}_y = \frac{2}{(e^Ty)^2}Qy - 2\frac{y^TQy}{(e^Ty)^3}e.$$

Now, because $y = \hat{y} + x^+ - x^-$, it follows that

$$g^+ = \bar{V}_y \quad \text{and} \quad g^- = -\bar{V}_y.$$

Exercises

1. How would the formulation of the re-balancing problem change if different transaction costs were associated with each asset?

2. Formulate the re-balancing problem for the case when the objective is to maximize return for a given value of risk.

3. Show that the denominator of (22.1.6) can be written as $(1 + e^T(x^+ - x^-))^2$. How would this affect the calculation of the gradient ∇F?

Solutions of the re-balancing problem

The program `sample7` can be used to compute solutions to the re-balancing problem outlined above. As an example, we consider the dataset `Real-75-5` and suppose that we already have a portfolio defined by the invested fractions

$$\hat{y}_1 = 0.152,\ \hat{y}_2 = 0.249,\ \hat{y}_3 = 0.309,\ \hat{y}_4 = 0.061,\ \hat{y}_5 = 0.229. \qquad (22.1.10)$$

(These values represent the minimum-risk portfolio, calculated on the basis of dataset `Real-50-5`, for a target return 0.2%.) When the extra twenty-five days' information in `Real-75-5` is taken into account, the portfolio (22.1.10) turns out to have an expected return of only 0.187% for a risk $V \approx 0.763$. The re-balancing problem seeks a new minimum-risk portfolio whose expected return is the original target value of 0.2%. We suppose that the transaction cost factors c_b and c_s are both 0.05.

Starting from the guessed solution $x^+ = x^- = 0$, `AL-SQP` solves the re-balancing problem in 9 iterations and 12 function calls and finds

$$x^+ = (0, 0.0958, 0, 0.0851, 0) \quad x^- = (0.0353, 0, 0.0416, 0, 0.1229).$$

Hence the re-balanced portfolio is given by

$$y_1 = 0.117,\ y_2 = 0.345,\ y_3 = 0.267,\ y_4 = 0.146,\ y_5 = 0.106 \qquad (22.1.11)$$

which has an expected return 0.2% for a risk $V \approx 0.654$.

It is interesting to compare the solution (22.1.11) with the optimal portfolio that would have been calculated if there were no transaction costs, namely

$$y_1 = 0.125,\ y_2 = 0.355,\ y_3 = 0.264,\ y_4 = 0.149,\ y_5 = 0.107. \qquad (22.1.12)$$

This produces an expected return of 0.2% with a risk $V \approx 0.651$.

We can also compare (22.1.11) with the new portfolio that would be set up from scratch – i.e., *not* re-balanced from an existing one – in order to give minimum risk for a return 0.2% when $c_b = 0.05$. This is

$$y_1 = 0.102,\ y_2 = 0.336,\ y_3 = 0.267,\ y_4 = 0.143,\ y_5 = 0.102 \qquad (22.1.13)$$

giving $V \approx 0.659$. This shows that, when there are transaction costs, an optimally re-balanced portfolio may be different from an optimal new one, even when both are based on the same data.

We now consider a second example. In this case the dataset is still `Real-75-5`, but the current portfolio has invested fractions

$$\hat{y}_1 = 0.3,\ \hat{y}_2 = 0.1,\ \hat{y}_3 = 0.1,\ \hat{y}_4 = 0.1,\ \hat{y}_5 = 0.4$$

giving an expected return 0.075%. We solve the re-balancing problem to minimize the portfolio risk for an expected return 0.1% and obtain

$$x^+ \approx (0.4,\ 1.04,\ 0.6,\ 0.53, 0.83) \quad x^- \approx (0.57,\ 0.92,\ 0.58, 0.57,\ 1.16).$$

This means that the re-balancing involves both buying and selling for each asset. The net effect is to give a new portfolio with

$$y_1 = 0.126,\ y_2 = 0.229,\ y_3 = 0.123,\ y_4 = 0.092,\ y_5 = 0.068. \qquad (22.1.14)$$

We might have assumed that the re-balancing problem as formulated in the previous section would automatically have given solutions in which *either* $x_i = 0$ *or* $x_{n+i} = 0$ – that is, where holdings of the i-th asset are changed either by selling or buying (but not both). However, we can now see that this, apparently reasonable, assumption is in fact false. Therefore we can add the extra *complementarity* constraints

$$x_i x_{n+i} = 0, \quad i = 1, .., n \qquad (22.1.15)$$

to the problem. Minimizing (22.1.6) subject to the extended constraint set (22.1.5), (22.1.7)–(22.1.9) and (22.1.15) gives, instead of (22.1.14), the re-balanced portfolio

$$y_1 \approx 0.279,\ y_2 \approx 0.355,\ y_3 \approx 0.0998,\ y_4 \approx 0.135,\ y_5 \approx 0.1. \qquad (22.1.16)$$

This gives the target return 0.1% for a risk $V \approx 0.659$. The original solution (22.1.14) corresponds to a risk $V \approx 0.635$. Therefore the solution involving both buying and selling of assets is actually better. This kind of, rather unexpected, behaviour is noted and discussed in [52].

Although the previous example is quite interesting, it arises in the context of a rather artificial problem. For the dataset `Real-75-5`, the least possible value of portfolio risk is $V_{min} \approx 0.635$ which corresponds to an expected portfolio return 0.157%. Therefore, in choosing to solve the minimum-risk re-balancing problem for a target return $R_p = 0.1\%$, we are opting for *worse* performance than that provided by the least-risky portfolio. If we re-balance using $R_p = 0.2\%$ (which exceeds the least-risky return) then the solution of the problem *without* the complementarity constraint turns out automatically to satisfy (22.1.15).

Exercises (To be solved using `sample7` or other suitable software.)
1. How does the re-balanced solution (22.1.11) for dataset `Real-75-5` compare with what would be given by the approach in section 2 of chapter 16?

2. How many iterations are needed by `IPM` and `SOLVER` to find solution (22.1.11)?

3. Suppose we are considering a portfolio with invested fractions $\hat{y}_1, .., \hat{y}_n$ with $e^T\hat{y} < 1$ (e.g., a portfolio that has already been re-balanced at least once). What changes would be needed in formulating a problem equivalent to minimizing (22.1.6) subject to constraints (22.1.5), (22.1.7), (22.1.8), (22.1.9)?

4. Use the data for problem T12 and re-balance the portfolio defined by $\hat{y}_1 = .. = \hat{y}_{10} = 0.1$ so as to obtain expected return 0.1% with minimum risk, assuming that $c_b = 0.02$ and $c_s = 0.01$. How does the result compare with that given by the minimum-change approach in section 2 of chapter 16?

5. Use data from problem T14 to re-balance the portfolio calculated in question 4 so as to maximize return for a risk $V_a = 0.025$. (Assume the cost factors are $c_b = c_s = 0.03$.)

2. Downside risk

It is easy to show (Exercise 1, below) that problem **Minrisk1** can be written as

$$\text{Minimize } V = \frac{1}{m}\sum_{j=1}^{m}[\sum_{i=1}^{n} r_{ij}y_i - R_p]^2 \qquad (22.2.1)$$

$$\text{subject to } \frac{1}{R_p}(\bar{r}^T y - R_p) = 0, \quad e^T y - 1 = 0. \qquad (22.2.2)$$

This representation shows that the calulation of risk V involves finding the portfolio return for each of the m time periods in the asset history and hence obtaining the total squared deviation from the target return. Let us define

$$R_j = \sum_{i-1}^{n} r_{ij} y_i \tag{22.2.3}$$

as the portfolio return for the j-th time period. Then it is clear that V in (22.2.1) makes no distinction between the two cases $R_j > R_p$ and $R_j < R_p$. Intuitively, however, we would probably regard the former as being a "good" situation while the latter is "bad". The term *downside* risk refers to methods of calculating risk so as to reflect the possibility of return being "too low" rather than "too high". Several definitions for downside risk have been proposed [61]. One approach mentioned in [61] and [62] is to write

$$V_d = \frac{1}{m} \sum_{j=1}^{m} [\min(0, R_j - R_p)]^2 \tag{22.2.4}$$

so that downside risk V_d only includes the squared deviations when portfolio expected return falls below target. Hence we could obtain an optimal portfolio, somewhat different from those in previous chapters, by finding invested fractions to minimize (22.2.4) subject to the constraints (22.2.2).

We can also form a downside risk version of problem **Minrisk0**. **Minrisk0** itself can be written in the form

$$\text{Minimize } \frac{1}{m} \sum_{j=1}^{m} [R_j - \bar{r}^T y]^2 \tag{22.2.5}$$

$$\text{subject to } e^T y = 1. \tag{22.2.6}$$

As in (22.2.4), we can replace the objective function in (22.2.5) by

$$V_d = \frac{1}{m} \sum_{j=1}^{m} [\min(0, R_j - \bar{r}^T y)]^2. \tag{22.2.7}$$

Exercises

1. Show that (22.2.1), (22.2.2) has the same solution as **Minrisk1**. (**Hint** consider section 6 of chapter 1.)

2. How could **Maxret1** be modified to maximize return for a given level of downside risk?

3. How would downside risk definitions be modified in a portfolio problem involving the constraint $e^T y \leq 1$?

Downside risk solutions

We consider problem T1 which uses the asset data in Table 4.1. With the standard definition of risk, the solution to this problem has invested fractions

$$y_1 \approx 0.42,\ y_2 \approx 0.34,\ y_3 \approx 0.01,\ y_4 \approx 0.2,\ y_5 \approx 0.04.$$

Now we seek the portfolio giving the same expected return $R_p = 1.15\%$ as for problem T1 but with minimum downside risk as defined by (22.2.4). Using SOLVER, we find that the resulting invested fractions are

$$y_1 \approx 0.32,\ y_2 \approx 0.32,\ y_3 \approx 0.1,\ y_4 \approx 0.26,\ y_5 \approx 0.$$

The downside risk value is $V_d \approx 0.00174$ while the standard risk $V \approx 0.00359$. This compares with $V_d \approx 0.00178$ and $V \approx 0.00344$ at the solution to the original problem T1. There is an appreciable alteration in the optimal invested fractions even though the actual value of downside risk has not changed much.

If we solve **Minrisk0** using the data in Table 4.1 we find that the least possible value for the standard risk function is $V \approx 0.001$ when

$$y_1 \approx 0.59,\ y_2 \approx 0.25,\ y_3 \approx -0.01,\ y_4 \approx -0.18,\ y_5 \approx 0.34.$$

This gives an expected portfolio return of about 1.085% and the downside risk value from (22.2.7) is about 0.0005. If we adjust the y_i so as to minimize (22.2.7) (subject, of course, to $e^T y = 1$) we get a very different solution

$$y_1 \approx 0.49,\ y_2 \approx 0.16,\ y_3 \approx 0.005,\ y_4 \approx 0.017,\ y_4 \approx 0.32.$$

This gives expected portfolio return about 1.086% and downside risk ≈ 0.00043.

Exercises

1. Plot the efficient frontier for the data in Table 4.1 for $1.05\% < R_p < 1.2\%$ using downside risk (22.2.4) in place of $V = y^T Qy$.

2. Consider the data for problem T3 and find the least-risky portfolios *(a)* using the standard definition of risk and *(b)* using downside risk (22.2.7).

3. For the data of problem T3, find the portfolio which gives expected return 1.27% while minimizing downside risk.

4. For the data in Table 4.1 formulate and solve the problem of finding the maximum return that can be obtained with a downside risk value of 0.0005.

3. Worst-case analysis

Following on from (22.2.1), we observe that another way to assess risk in terms of deviation from some target return would be to define

$$V_w = Max_{1 \le j \le m} |R_j - R_p| \tag{22.3.1}$$

where R_j is defined in (22.2.3) as the portfolio return in time period j of the historical asset data. Here, rather than taking all time periods into account, we are considering only the biggest departure from target performance. This is an example of *worst-case* analysis.

Yet another version of the minimum-risk portfolio selection problem would involve minimizing V_w subject to the familiar constraints (22.2.2). In practice, to avoid dealing directly with the non-smooth function (22.3.1), we would formulate the problem in the same way as in minimax data-fitting. That is, using $y_1, .., y_n$ and z as optimization variables,

$$\text{Minimize} \quad z \tag{22.3.2}$$

$$\text{subject to} \quad -z \le R_j - R_p \le z, \quad \text{for} \quad j = 1, .., m \tag{22.3.3}$$

$$\text{and subject to} \quad \frac{1}{R_p}(\bar{r}^T y - R_p) = 0, \quad e^T y - 1 = 0. \tag{22.3.4}$$

Clearly we could also adopt a worst-case view of downside risk by considering

$$\text{Minimize} \quad z \tag{22.3.5}$$

$$\text{subject to} \quad -z \le \min(0, R_j - R_p) \quad \text{for} \quad j = 1, .., m \tag{22.3.6}$$

$$\text{and subject to} \quad \frac{1}{R_p}(\bar{r}^T y - R_p) = 0, \quad e^T y - 1 = 0. \tag{22.3.7}$$

Optimization based on worst-case analysis is a much wider topic than we can consider here. A comprehensive study is given by Rustem and Howe [63]. A characteristc feature of the approach is that it requires the solution of minimax problems. An example, given in [63] shows how we can find an optimal portfolio taking account of different models or scenarios. For instance, it could be argued that investment strategies should be based chiefly on *recent* asset performance as captured in a mean-return vector $\bar{r}^{(1)}$ and variance-covariance matrix $Q^{(1)}$ computed from data for m_1 days where m_1 is quite small. On the other hand, it might be thought unwise to ignore longer term trends contained in $\bar{r}^{(2)}$ and $Q^{(2)}$ derived from a history of m_2 days, where $m_2 > m_1$. It is possible to respect *both* points of view by solving

$$\text{Minimize} \quad Max(y^T Q^{(1)} y, y^T Q^{(2)} y) \tag{22.3.8}$$

subject to

$$\frac{1}{R_p}(\bar{r}^{(1)T} y - R_p) = 0, \quad \frac{1}{R_p}(\bar{r}^{(2)T} y - R_p) = 0, \quad e^T y - 1 = 0. \tag{22.3.9}$$

In practice, this problem can be posed using the extra variable z and inequality constraints – as in (22.3.2), (22.3.3), (22.3.4) – to avoid dealing with the non-smooth objective function (22.3.8). Such an approach can always be applied to *discrete* minimax problems, where the maximization part involves a (possibly small) number of distinct alternatives. It sometimes happens, however, that worst-case optimization involves continuous maximization of a function as in the general formulation

$$\text{Minimize (w.r.t. } u\text{) [Max (w.r.t. } v\text{) } F(u,v)] \tag{22.3.10}$$

subject to

$$c_i(u) \geq 0, \;\; i = 1,..,m_u; \;\; c_i(v) \geq 0, \;\; i = m_u + 1,..,m_u + m_v. \tag{22.3.11}$$

Formulation and solution of such problems is discussed in [63].

Exercises

1. Formulate (22.3.8), (22.3.9) so that it can be solved using an algorithm for solving continuously differentiable inequality constrained problems.

2. Formulate a maximum-return problem which uses a worst-case definition of portfolio risk.

Worst-case portfolio solutions

Our example is based on problem T1 and we use SOLVER to calculate results. If risk is defined as V_w from (22.3.1) then the minimum-risk portfolio giving expected return 1.15% from the assets in Table 4.1 has invested fractions

$$y_1 \approx 0.77, \; y_2 \approx 0.48, \; y_3 \approx 0.08, \; y_4 \approx -0.4, \; y_5 \approx 0.07.$$

The optimal value of V_w is about 0.087. The change in risk definition has caused a substantial change from the optimal portfolio (4.4.6) given by **Minrisk1** with $V = y^T Q y$. If we add the inequality constraints $y_i \geq 0, \; i = 1,..,5$ to

(22.3.2) - (22.3.4) to prevent short-selling then SOLVER gives

$$y_1 \approx 0.45,\ y_2 \approx 0.51,\ y_3 \approx 0.02,\ y_4 \approx 0,\ y_5 \approx 0.03$$

with $V_w \approx 0.095$.

We can use the data in Table 4.1 to give an instance of problem (22.3.8), (22.3.9). We let $\bar{r}^{(1)}$ be the mean returns based on periods 5-10. $Q^{(1)}$ is the corresponding variance-covariance matrix. Values for $\bar{r}^{(2)}$ and $Q^{(2)}$ are calculated from periods 1-10. When (22.3.8), (22.3.9) is formulated as in question 1 of the exercises above (with $R_p = 1.15\%$) SOLVER finds the optimal portfolio

$$y_1 \approx -0.03,\ y_2 \approx 0.28,\ y_3 \approx 0.29,\ y_4 \approx 0.55,\ y_5 \approx -0.1.$$

It is helpful to consider the Lagrange multipliers at this computed solution. The multiplier associated with the constraint

$$\frac{1}{R_p}(\bar{r}^{(1)T}y - R_p) = 0$$

is negative, indicating that the objective function could be reduced if the expected return were allowed to exceed the target R_p rather than matching it exactly. If the problem is reformulated with inequality constraints

$$\frac{1}{R_p}(\bar{r}^{(1)T}y - R_p) \geq 0, \quad \frac{1}{R_p}(\bar{r}^{(2)T}y - R_p) \geq 0$$

then the solution is the same as (4.4.6) for the original problem T1.

Exercises

1. Use asset data Table 4.1 and problem (22.3.5)–(22.3.7) to find the portfolio which gives target return 1.15% for the least value of worst-case downside risk.

2. For the data in Table 4.1, compare the efficient frontier obtained using V_w as a measure of risk with that obtained using the standard definition $V = y^T Q y$.

3. Using the returns in dataset `Real-20-5` (listed in question 3 of the Exercises in section 6 of chapter 11) compute mean returns $\bar{r}$ and variance-covariance matrix Q based on data for the most recent 10, 15 and 20 trading days. Hence set up and solve a problem similar to (22.3.8), (22.3.9) to give a worst-case optimal portfolio having a specified expected return.

Chapter 23

GLOBAL UNCONSTRAINED OPTIMIZATION

1. Introduction

Figure 23.1 is a plot of the contours of the function

$$F(x) = (\bar{r}^T x - 1.22)^2 + 1000(x^T Q x - 0.00375)^2 \tag{23.1.1}$$

where $\bar{r}$ and Q are obtained from the asset data in Table 1.1. Minimizing this function can be regarded as an attempt to get as close as possible to meeting requirements on both return and risk. The contours show that that (23.1.1) has *two* local minima. These lie near to (0.65, 0.15) and (0.6, 0.55) and are separated by a maximum near (0,6, 0.3).

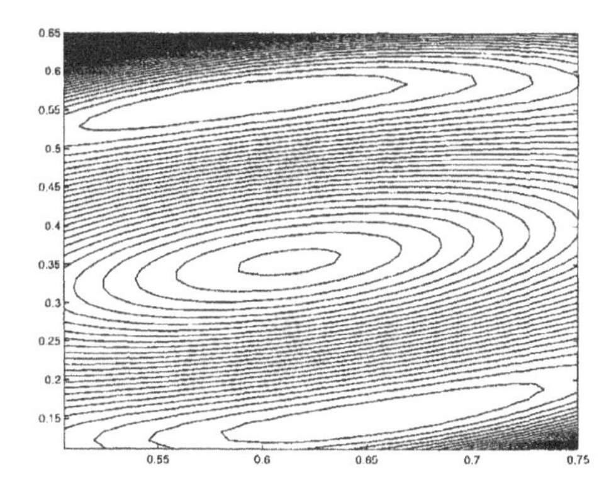

Figure 23.1. Contours of (23.1.1)

In practical optimization problems it is quite common for several local solutions to exist. We shall outline below some optimization methods which seek global, rather than local minima. Unfortunately there are no conditions that can be evaluated at a point x^* which will establish whether or not it is a global optimum. Hence the methods we describe cannot in general *guarantee* success – at least, not within a practicable amount of computing effort. For a fuller account of the global optimization problem and its solution see [13].

2. Multi-start methods

A heuristic approach to the global minimization problem would be to run a local minimization algorithm from many different starting points and then pick the best of the solutions. This strategy can sometimes be effective, but its drawbacks are *(i)* it is wastefully expensive since many local searches may yield the same result; and *(ii)* it provides no assurance that the local optima found do actually include the global solution. The approach can be formalised and made more efficient by the incorporation of some statistical theory. To illustrate this we will mention two ideas which are used in a global optimization method proposed by Rinnooy-Kan & Timmer [64, 65].

Cluster analysis can be used to see if different local optimizations are tending to the same result. If we allow all the optimizations to perform a fixed number of iterations (not too large) we can estimate how many of the searches seem to be heading for different solutions. We would then perform another set of iterations of a (probably much smaller) number of optimization calculations and repeat the cluster analysis. Continuing in this way, we would expect to locate multiple minima more cheaply than with the basic “scattergun” approach.

Bayesian estimation can be used to determine an expected number of minima of the objective function on the basis of the number that have been found so far (W) and the number of local searches used (N_s). A formula for the expected total number of local solutions is

$$W_t = \frac{W(N_s - 1)}{N_s - W - 2}.$$

Thus, if 5 minima are found in 30 searches $W_t \approx 6.3$, which suggests that some solutions may not yet have been found. If no more minima are found when 100 searches have been completed then $W_t \approx 5.3$ and it is now more reasonable to suppose that there are only five local solutions.

Rinnooy Kan and Timmer have developed a global optimization algorithm using both these ideas. An initial iteration (using N_s starting points and clustering) produces W local optima, say. If $W_t >> W$ then further cycles of local optimization and clustering are performed from new starting points until $W_t < W + 0.5$ (say). The algorithm of Rinnooy Kan and Timmer also includes strategies, not described here, to ensure that additional starting points are not chosen too close to minima that have already been found or to starting points that have been used previously.

3. DIRECT

In practice, most global minimizers are applied in some restricted region of variable-space, typically in a "hyperbox" defined by

$$l_i \leq x_i \leq u_i.$$

The algorithm DIRECT (Jones et al [6]) relies on the use of such rectangular bounds on the area of search. It works by systematic exploration of sub-boxes in the region of interest. In the limit, as the number of iterations becomes infinite, it will sample the whole region and, in that sense, the algorithm is guaranteed to converge. The practical performance of the method depends on the way in which it chooses which sub-boxes to explore first, because this will determine whether the global minimum can be approximated in an acceptable number of iterations.

To describe the method we consider first the one-variable problem of finding the global minimum of $F(x)$ for $0 \leq x \leq 1$. (Any problem can be reduced to this range by a simple transformation.) We begin by dividing $[0,1]$ into three equal sub-ranges and evaluating the function at their midpoints. We then identify the range with the least function value and trisect it, evaluating F at the midpoints of the new ranges. We then have a situation like that shown in Figure 23.2.

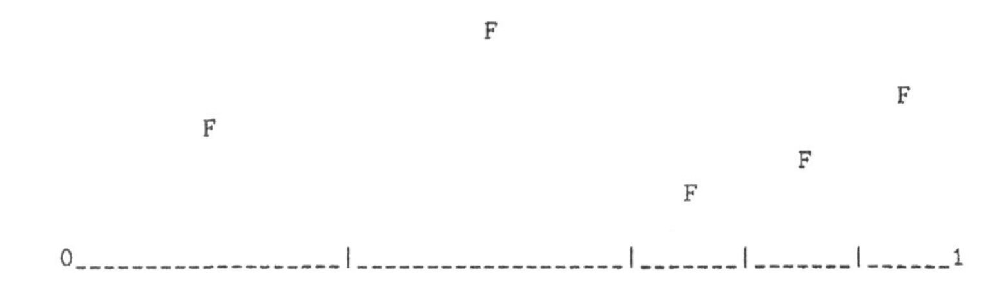

Figure 23.2. One iteration of DIRECT on a one-variable problem

There are now trial ranges of two different widths, namely, $\frac{1}{3}$ and $\frac{1}{9}$. For *each* of these widths we trisect the one with the smallest value of F at the centre. This is depicted in Figure 23.3.

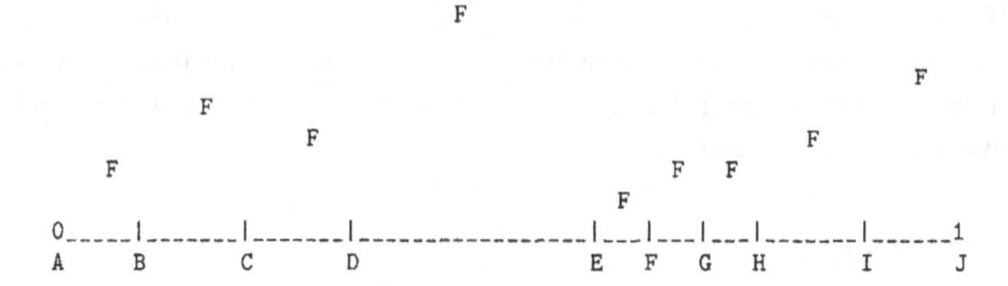

Figure 23.3. Two iterations of DIRECT on a one-variable problem

On the next iteration we have *three* candidate range-sizes, $\frac{1}{3}$, $\frac{1}{9}$ and $\frac{1}{27}$. For each of them we can trisect the representative with smallest F-value at its centre. In the diagram, this means we would subdivide the intervals DE, AB and EF. Continuing in this way, we can systematically explore the whole range in a way that concentrates on the most promising regions first. Thus we aim to find a good estimate of the global optimum before the iteration count gets too high.

The basic idea just outlined can be made more efficient if we refine the definition of a "promising" range. Suppose that $d_1,..,d_p$ are the p different range-sizes at the start of an iteration. Suppose also that F_j denotes the smallest of all the function values at the centres of ranges of width d_j. We will trisect the range containing F_j *only* if a "potential optimality" test is satisfied. The argument behind this test is based upon *Lipschitz constants*, which are bounds on the magnitude of the first derivative of F. If F has Lipschitz constant L then, within the range containing F_j, we have the bounds

$$F_j + \frac{1}{2}Ld_j \geq F(x) \geq F_j - \frac{1}{2}Ld_j.$$

We do not normally know a Lipschitz constant for F. However, the range containing F_j can be said to be potentially optimal if *there exists* a Lipschitz constant L such that

$$F_j - \frac{1}{2}Ld_j < F_i - \frac{1}{2}Ld_i \quad \text{for } i = 1,..,p; \quad i \neq j. \tag{23.3.1}$$

For this to be true we need

$$L > 2 \times max\{\frac{F_j - F_i}{d_j - d_i}\} \quad \text{for all } i : d_i < d_j$$

and

$$L < 2 \times min\{\frac{F_j - F_i}{d_i - d_j}\} \quad \text{for all } i : d_i > d_j$$

If these conditions on L are inconsistent then the range containing F_j cannot be considered potentially optimal and hence it need not be subdivided. This consideration can save wastefully expensive function evaluations when there are many different candidate ranges.

Another "filter" can be used to reduce the number of ranges to be subdivided. If F_{min} is the smallest function value found so far then the range containing F_j will not be trisected unless there exists an L which satisfies (23.3.1) and also

$$F_j - \frac{1}{2}Ld_j < F_{min} - \varepsilon|F_{min}| \tag{23.3.2}$$

where ε is a user specified parameter (typically about 0.001). Condition (23.3.2) suggests that subdivision of the range containing F_j can be expected to produce a non-trivial improvement in the best function value so far.

The ideas outlined so far can be extended to problems in several variables [6]. The original search region becomes a hyperbox rather than a line segment and the initial subdivision is into three hyperboxes by trisection along the longest edge. The objective function is evaluated at the centre point of each of these boxes; and the size of each box is taken as the length of its diagonal. The box with the smallest value of F at its centre is subdivided by trisection along its longest side and the process of identification and subdivision of potentially optimal hyperboxes then continues as in in the one-variable case. (There are refinements for dealing with the subdivison of boxes which have several longest sides [6].)

DIRECT can get good estimates of global optima quite quickly. Since it only uses function values, it can be applied to non-smooth problems or to those where the computation of derivatives is difficult. One drawback, however, is that there is no hard-and-fast convergence test for stopping the algorithm. One can simply let it run for a fixed number of iterations or else choose to terminate if there is no improvement in the best function value after a prescribed number of function evaluations. Neither of these strategies, however, will *guarantee* that enough iterations have been done to identify the neighbourhood of the global optimum.

4. Numerical examples

A variation on the problem of minimizing the function (23.1.1) is to minimize

$$F(x) = (\bar{r}^T x - R_p)^2 + 1000(x^T Qx - V_a)^2 + \rho\psi(x) \tag{23.4.1}$$

where

$$\psi(x) = \begin{cases} (x_2 - 0.5)^2 & \text{if } x_2 > 0.5 \\ 0 & \text{otherwise} \end{cases} \tag{23.4.2}$$

Like (23.1.1), this function will have two local minima (at least, when ρ is small) which lie in the vicinity of (0.65,0.15) and (0.6,0.55). However, the extra term $\rho\psi(x)$ will mean that F has different values at these two minima; and, for $\rho > 0$, it will be the one near (0.65, 0.15) for which F is smaller.

The program `sample13` minimizes (23.4.1) using first a quasi-Newton method and then DIRECT. We shall run `sample13` with data from Table 1.1, taking $R_p = 1.22$, $V_a = 0.00375$ and $\rho = 0.1$. From the starting guess $x = (0.5, 0.5)^T$, `QNw` converges to

$$x_1 \approx 0.549, \quad x_2 \approx 0.540 \quad \text{with } F \approx 2.6 \times 10^{-4}.$$

But, when started from $x = (0,0)^T$, it converges to

$$x_1 \approx 0.658 \quad x_2 \approx 0.152 \quad \text{with } F \approx 0.$$

Hence the first calculation has terminated at a local optimum, rather than the global optimum found by the second calculation. This is typical of the behaviour of a local minimization method: it usually converges to the local minimum "nearest to" the starting guess. If we run `QNw` again from ten random starting points with $0 < x_i < 1$ (thus imitating a weak form of multi-start) it finds the global optimum on seven occasions.

We now use DIRECT to solve the same problem with $0 < x_1, \ x_2 < 1$ as the search region. After 10 iterations the best point found is

$$x_1 \approx 0.661 \quad x_2 \approx 0.154 \quad \text{with } F \approx 1.8 \times 10^{-6}$$

which is quite a good estimate of the global solution. The global optimum is found to full accuracy in about 20 iterations.

If we use $\rho = 1$ as the weighting parameter in (23.4.1) then in twelve starts of `QNw` we get the global solution five times. There are now *two* other stopping points, one near (0.81,0.5) and one near (0.5,0.5). Thus, when we run DIRECT in the box $0 < x_i < 1$, the starting point is very close to a local solution. Nevertheless, DIRECT still finds the global optimum in about 20 iterations.

Exercise
Run `sample13` with data from Table 1.2 and using $R_p = 0.075, V_a = 0.0015$ and determine the global and local minima when $\rho = 0.1$. Estimate the size of the *region of attraction* for the quasi-Newton method around each minimum. Show that DIRECT succeeds in finding the global minimum whichever of these regions it is started in. Investigate the importance of search-box size in ensuring that DIRECT finds the global optimum.

5. Global optimization in portfolio selection

In all the examples so far we have allowed the invested fractions y_i to take any values between some upper and lower bounds. We now consider a slightly more complicated situation in which we do not want to invest *very small* amounts in any asset. For instance, suppose we want each y_i to satisfy

$$\text{either } y_i = 0 \quad \text{or} \quad y_i \geq y_{min}. \tag{23.5.1}$$

This kind of *buy-in threshold* constraint is discussed in [53]. It states that if we buy an asset we must invest at least y_{min}% of our capital in it.

Minimum risk with buy-in threshold

To solve the minimum-risk problem subject to a restriction of the form (23.5.1) we can use an extension of **Minrisk2u**. The optimization variables are $x_1,...,x_n$ and the invested fractions are $y_i = x_i^2$, $i = 1,..,n$. Then we can minimize

$$F = y^T Qy + \rho(e^T y - 1)^2 + \rho(\frac{\bar{r}^T y}{R_p} - 1)^2 + \mu \sum_{i=1}^{n} \psi(y_i)^2 \tag{23.5.2}$$

where

$$\psi(y_i) = \begin{cases} 0 & \text{if } y_i \geq y_{min} \\ 4y_i(y_{min} - y_i)/y_{min}^2 & \text{if } 0 \leq y_i < y_{min} \end{cases} \tag{23.5.3}$$

The function $\psi(y_i)$ takes values between zero and one and it is used to penalise any y-values which lie in the forbidden zone $0 < y_i < y_{min}$. Since the function (23.5.2) is likely to have several local minima – each corresponding to some of the y_i being close to zero or close to y_{min} – we need to approach it using a global optimization technique. The program `sample14` does this. It first solves a minimum-risk problem (without short-selling) based upon artificial or real-life data for asset histories. Then, on the basis of the computed solution, the user can specify a value for y_{min} and the program seeks to minimize the function (23.5.2) The following example is based on the dataset `Real-100-5`. The solution to **Minrisk2u** when $R_p = 0.25\%$ has invested fractions

$$y_1 \approx 0.132, \; y_2 \approx 0.368, \; y_3 \approx 0.345, \; y_4 \approx 0.117, \; y_5 \approx 0.037.$$

The portfolio risk is $V \approx 0.6894$. The investment in asset 5 is less than 5% of the total and so we choose $y_{min} = 0.05$ in (23.5.2) and set the penalty parameter $\mu = 1$. At the minimum of (23.5.2) we expect *either* that y_5 will be near zero *or* that $y_5 \approx 0.05$. That is, we expect a change of about ± 0.03 in y_5. In order to maintain the total investment $\sum y_i = 1$, this means there could be a compensating change of up to ± 0.03 in any of the other invested fractions. Therefore we seek the *global* optimum of (23.5.2) in the hyperbox

$$0.1 \leq y_1 \leq 0.16;\ 0.34 \leq y_2 \leq 0.4;\ 0.31 \leq y_3 \leq 0.37; \qquad (23.5.4)$$

$$0.09 \leq y_4 \leq 0.15;\ 0 \leq y_5 \leq 0.06. \qquad (23.5.5)$$

When started from the midpoint of this region, where (23.5.2) has a value of about 2.49, QNw converges to a minimum with

$$y_1 \approx 0.152;\ y_2 \approx 0.381;\ y_3 \approx 0.348;\ y_4 \approx 0.118;\ y_5 \approx 0 \qquad (23.5.6)$$

where (23.5.2) has a value approximately 0.7005. The portfolio risk is 0.7001 and the return is 0.25%, as required.

We now apply DIRECT to (23.5.2) within the hyperbox (23.5.4), (23.5.5). After 10 iterations it gives a point where (23.5.2) has a value of about 0.6949. This is appreciably lower than the optimum found by the quasi-Newton method. The invested fractions are

$$y_1 \approx 0.122;\ y_2 \approx 0.369;\ y_3 \approx 0.339;\ y_4 \approx 0.111;\ y_5 \approx 0.058 \qquad (23.5.7)$$

with a portfolio risk of 0.6935. This approximate solution has $y_5 \approx 0.05$ which suggests that (23.5.6) (where $y_5 \approx 0$) is only a local optimum.

We note that DIRECT may not find optima to very high accuracy. This is partly because it does not use derivatives and partly because it only samples function values at the centres of hyperboxes. A common strategy, therefore, is to run a quasi-Newton method again from the best point located by DIRECT, in order to refine the approximate solution. If we do this, we find that an accurate global minimum (23.5.2) has a value of about 0.691 and the invested fractions are

$$y_1 \approx 0.125;\ y_2 \approx 0.364;\ y_3 \approx 0.344;\ y_4 \approx 0.116;\ y_5 \approx 0.05. \qquad (23.5.8)$$

Here the target return 0.25% is still achieved and the portfolio risk is 0.6906. This is only slightly worse than the minimum risk obtained when the constraint ($y_i = 0$ or $y_i \geq 0.05$) was not applied.

Maximum return with buy-in threshold

We can solve maximum return problems with the buy-in constraint (23.5.1) by minimizing the function

$$F = -\bar{r}^T y + \rho(e^T y - 1)^2 + \rho(\frac{y^T Q y}{V_a} - 1)^2 + \mu \sum_{i=1}^{n} \psi(y_i)^2. \qquad (23.5.9)$$

The program `sample14` deals with this in the same way as the minimum risk problem involving (23.5.2). As an example we use the same dataset `Real-100-5` and take the acceptable risk as $V_a = 0.75$. With $\rho = 10^3$ the solution of **Maxret2u** is

$$y = (0.0462,\ 0.4022,\ 0.4162,\ 0.1053,\ 0.0299)^T,$$

giving a portfolio return of 0.2877%. To make all non-zero invested fractions greater than or equal to 0.05 we minimize (23.5.9) with $y_{min} = 0.05$ and $\mu = 1$. A reasonable box to search in is

$$0 \le y_1 \le 0.06;\ 0.36 \le y_2 \le 0.44;\ 0.38 \le y_3 \le 0.46;$$

$$0.06 \le y_4 \le 0.12;\ 0 \le y_5 \le 0.06.$$

When started from the middle of this region, `QNw` finds a minimum at

$$y \approx (0.0034,\ 0.4037,\ 0.4112,\ 0.1529,\ 0.0030)^T$$

where (23.5.9) has a value of about 0.6472. Twenty iterations of DIRECT in the same box produces the much lower value, $F = -0.276$, at

$$y \approx (0.0578,\ 0.399,\ 0.419,\ 0.0746,\ 0.0494)^T.$$

This indicates that it is better for y_1 and y_5 to be near 0.05 rather than near zero, as in the quasi-Newton *local* solution. If we use `QNw` to refine the approximate global minimum given by DIRECT we get the accurate result

$$y_1 = 0.05;\ y_2 \approx 0.396;\ y_3 \approx 0.417;\ y_4 \approx 0.0869;\ y_5 = 0.05$$

where the portfolio return is about 0.2855%, which is only slightly worse than was possible when there was no lower limit on the y_i.

Exercises (To be solved with `sample14` or other suitable software)

1. Obtain expressions for the gradient and Hessian of the function ψ in (23.5.3) *remembering that the independent variables are* $x_1, ..., x_n$.

2. Solve problem T12a with extra constraints $y_i = 0$ or $y_i \ge 0.04$ for $i = 1, .., 10$.

3. Solve problem T14a with extra constraints $y_i = 0$ or $y_i \ge 0.03$ for $i = 1, .., 10$.

4. Modify problem (23.5.2) so as to allow short-selling and incorporate an extra constraint of the form

$$y_i = 0 \quad \text{or} \quad y_i \ge y_{min} \quad \text{or} \quad y_i \le -y_{min}.$$

Hence seek a solution of problem T12 with $y_{min} = 0.02$.

5. Apply the approach of this section to the *roundlot* constraint [53].

Global solution

It's the only place
to be – if you can get there.
(If not, you won't know.)

Appendix - Accessing SAMPO software

The SAMPO fortran90 codes can be obtained by anonymous ftp from the site ftp.feis.herts.ac.uk.
This means that you must give ftp as your username and set your email address as the password.

A typical login dialogue under Unix might be as follows (where underlining shows user response).

```
 ftp ftp.feis.herts.ac.uk
Connected to brian.feis.herts.ac.uk.
220 ProFTPD 1.2.9 Server (FEIS FTP Server) [brian.feis.herts.ac.uk]


Name (ftp.feis.herts.ac.uk:matqmb): ftp
331 Anonymous login ok, send your complete email address as your password.
Password: inewton@cam.ac.uk
```

Once attached, you must change directory to pub/matqmb where the *.f90 codes will be found. There are also some data files corresponding to the tables of asset return histories quoted in the text. A short READ.ME file contains further information.

The Sampo

Corn and gold and salt
(in no especial order)
ground out on demand:

a wondrous device
but (like the dot.com bubble)
much too good to last.

A fight for control:
The Sampo is lost at sea
and nobody wins.

That's why the ocean's
salt. And why all beaches are
strewn with corn and gold.

References

1. R.J. Vanderbei *Linear Programming: Foundations and Extensions*, Kluwer Academic Publishers, 1996.

2. H.M. Markowitz, Portfolio Selection, Journal of Finance, March 1952, pp 77-91.

3. H.M. Markowitz *Portfolio Selection: Efficient Diversification of Investments* John Wiley (1959). Second Edition Blackwell (1991).

4. R. Hooke and T.A. Jeeves, Direct Search Solution of Numerical and Statistical Problems, J.ACM **8**, pp 212-229, 1961.

5. J.A. Nelder and R. Mead, A Simplex Method for Function Minimization, Comp. J. **7**, pp 308-313, 1965.

6. D.R.Jones, C.D. Perttunen and B.E. Stuckman, Lipschitzian Optimization without the Lipschitz Constant, Journ. Opt. Theory & Applics, **79**, pp 157-181, 1993

7. Frontline Systems Inc, www.solver.com

8. Micosoft Corporation, www.microsoft.com

9. The Mathworks Inc., www.mathworks.com

10. P. Wolfe, Convergence Conditions for Ascent Methods, SIAM Review, **11**, pp 226-235, 1969.

11. L. Armijo, Minimization of Functions having Continuous Partial Derivatives, Pacific J. Maths, **16**, pp 1-3, 1966.

12. N.J. Higham, *Accuracy and Stability of Numerical Algorithms*, SIAM, Philadelphia, 1996.

13. C.A. Floudas and P.M. Pardolos eds, *Encyclopedia of Optimization*, Kluwer Academic Publishers, 2001.

14. A Griewank, *Evaluating Derivatives: Principles and Techniques of Automatic Differentiation*, SIAM, 2000.

15. P.E. Gill and W. Murray, Newton-type Methods for Unconstrained and Linearly Constrained Optimization, Mathematical Programming **30**, pp 176-195, 1974.

16. R. Schnabel and E. Eskow, A New Modified Cholesky Factorization, SIAM Journal on Scientific Computing **11**, pp 1136-1158, 1991.

17. A. Jennings and J.J. McKeown, *Matrix Computation*, Second Edition, John Wiley, 1992.

18. A.R. Conn, N.I.M. Gould and Ph. L. Toint *Trust Region Methods*, MPS-SIAM Series on Optimization, Philadelphia, 2000.

19. M.C. Bartholomew-Biggs, A Newton Method with a Two-dimensional Line Search, Advanced Modeling and Optimization, www.ici.ro/camo/journal **5** pp 223-245, 2003,

20. J.Z. Zhang and C.X. Xu, A Class of Indefinite Dog-leg Methods for Unconstrained Minimization, SIAM J. Optim. **9**, pp 646-667, 1999.

21. W.C. Davidon, Variable-metric Method for Minimization, AEC Reprt ANL5990, Argonne National Laboratory, 1959.

22. R. Fletcher and M.J.D. Powell, A Rapidly Convergent Descent Method for Minimization, Computer Journal **6**, pp 163-168, 1963

23. C.G. Broyden, The Convergence of a Class of Double Rank Minimization Algorithms, Part 1, Journ. Inst. Maths. Applics **6**, pp 76-90, 1970

24. C.G. Broyden, The Convergence of a Class of Double Rank Minimization Algorithms, Part 2, Journ. Inst. Maths. Applics **6**, pp 222-231, 1970.

25. L.C.W. Dixon, Quasi-Newton Algorithms Generate Identical Points, Math. Prog. **2**, pp 383-387, 1972

26. L.C.W. Dixon, Quasi-Newton Algorithms Generate Identical Points, Part 2 - Proofs of Four New Theorems, Math. Prog. **3**, pp 345-358, 1972

27. M.J.D. Powell, On the Convergence of the Variable-metric Algorithm, J. Inst. Maths. Applics **7**, pp 21-36, 1971

28. M.J.D. Powell, Some Global Convergence Properties of a Variable-metric Algorithm without Line Searches, R.W. Cottle & C.E. Lemke eds, *Nonlinear Programming* AMS, 1976

29. D. Pu, The Convergence of Broyden Algorithms without Convexity Assumption, System Science & Maths Science **10**, pp 289 - 298, 1997.

30. J.E. Dennis and R.B. Schnabel, *Numerical methods for Unconstrained Optimization and Nonlinear Equations*, Prentice-Hall, 1983.

31. M.R. Hestenes & E.L. Stiefel, Methods of Conjugate Gradients for Solving Linear Systems, J. Res. Nat. Bureau of Standards **49**, pp 409-436, 1952.

32. R. Fletcher & C. Reeves, Function Minimization by Conjugate Gradients, Comp. J. **7**, pp 149-154, 1964.

33. E. Polak & G. Ribiere, Note sur la Convergence de Methode de Directions Conjugees, Revue France Inform. Rech. Oper. **16**, pp 35-43, 1969.

34. C.G. Broyden and M.T. Vespucci, *Krylov Solvers for Linear Algebraic Systems*, Studies in Computational Mathematics **11**, Elsevier, 2004.

35. R. Dembo, S. Eisenstat and T. Steihaug, Inexact Newton Methods, SIAM Journal on Numerical Analysis **10**, pp 400-408, 1982.

36. K. Levenberg, A Method for the Solution of Certain Nonlinear problems in Least Squares, Quart. Appl. Maths., **2**, pp 164-168, 1944.

37. D.W. Marquardt, An Algorithm for Least Squares Estimation of Nonlinear parameters, SIAM J., **11**, pp 111-115, 1963.

38. G.E. Box and G.M. Jenkins, *Time-series Analysis for Forecasting and Control*, Holden Day, 1970.

39. M.C. Steinbach, Markowitz Revisited: Mean-variance Models in Financial Portfolio Analysis, SIAM Review **43**, pp 31-85, 2001.

40. M.J.D. Powell, A Fast Algorithm for Nonlinearly Constrained Optimization Calculations, G. Watson ed. *Numerical Analysis, Dundee 1977*, Vol 630 of Lecture Notes in Mathematics, Springer, 1978.

41. L.S. Lasdon, A.D. Waren, A.Jain and M. Ratner, Design and Testing of a Generalised Reduced Gradient Code for Nonlinear Programming, ACM Trans. Math. Soft. **4**, pp 34-50, 1978.

42. A.V. Fiacco and G.P. McCormick, *Nonlinear Programming - Sequential Unconstrained Minimization Techniques*, John Wiley, 1968. Reissued by SIAM Classics in Applied Mathematics, 1990.

43. M.J.D. Powell, A Method for Nonlinear Constraints in Minimization Problems, R.Fletcher ed *Optimization*, Academic Press, 1969.

44. R. Fletcher and S.A. Lill, A Class of Methods for Nonlinear programming II: Computational Experience. J.B. Rosen, O.L. Mangasarian and K. Ritter, eds *Nonlinear programming*, Academic Press, 1972.

45. R.B. Wilson, A Simplicial Method for Concave Programming, PhD Dissertation, Harvard University, Cambridge MA, 1963.

46. S.P. Han, Superlinearly Convergent Variable-metric Algorithms for General Nonlinear Programming Problems, Math. Prog., **11**, pp 263-282, 1976.

47. W. Murray An Algorithm for Constrained Optimization, R. Fletcher ed *Optimization*, Academic Press, 1969.

48. M.C. Biggs, Constrained Minimization using Recursive Equality Quadratic Programming, F.A. Lootsma ed *Numerical Methods in Nonlinear Optimization*, Academic Press, 1972.

49. M.C. Bartholomew-Biggs, Recursive Quadratic Programming Methods Based on the Augmented Lagrangian, Math. Prog. Study **31**, pp 21-41, 1987.

50. N. Maratos, *Exact Penalty Function Algorithms for Finite-dimensional and Control Optimization Problems*, PhD Thesis, London University, 1978

51. R. Fletcher and S. Leyffer, Nonlinear Programming without a Penalty Function, Numerical Analysis Report NA/171, Department of Mathematics, University of Dundee, 1997.

52. J. Mitchell and S. Braun, Rebalancing an Investment Portfolio in the Presence of Transaction Costs, http://www.rpi.edu/ mitchj/papers/transcosts.html, Rensselaer Polytechnic Institute, 2002.

53. N.J. Jobst, M.D. Horniman, C.A. Lucas and G. Mitra, Computational Aspects of Alternative Portfolio Selection Models in the Presence of Discrete Asset Choice Constraints, Qunatitative Finance **1**, pp 1-13, 2001.

54. R. Fletcher, A General Quadratic Programming Algorithm, J. Inst. Maths Applics **7**, pp 76-91, 1971

55. R.T. Rockafellar, A Dual Approach to Solving Nonlinear Programming Problems using Unconstrained Optimization, Math. Prog, **5**, pp 354-373, 1973.

56. N. Karmarkar, A New Polynomial Time Algorithm for Linear Programming, Combinatorica **4**, pp 373- 395, 1984.

57. S.J. Wright, Recent Developments in Interior Point Methods, M.J.D. Powell and S. Scholtes, eds, *System Modelling and Optimization: Methods, Theory and Applications*, Kluwer, 1999.

58. R.H. Byrd, M. Hribar and J. Nodedal, An Interior Point Algorithm for Large Scale Nonlinear programming, OTC Technical Report 97/05, Optimization Technology Center, 1997.

59. A. Forsgren and P.E. Gill, Primal-dual Interior Point Methods for Nonconvex Nonlinear Programming, SIAM J. Opt. **8** pp1132-1152, 1998.

60. N.I.M. Gould and Pl. L. Toint, SQP Methods for Large-scale Nonlinear Programming, M.J.D. Powell and S. Scholtes, eds, *System Modelling and Optimization: Methods, Theory and Applications*, Kluwer, 1999.

61. D. Nawrocki, A Brief History of Downside Risk Measures, Technical Report, Villanova University, Arcola, PA.

62. Computational and Empirical Study of Downside Risk and Internationally Diversified Portfolios, Department of Computing Report, Imperial College, London University, 2003.

63. B. Rustem and M. Howe, *Algorithms for Worst-case Design and Applications to Risk Management*, Princeton University press, 2002.

64. A. Rinnooy-Kan and G. Timmer, Stochastic Global Optimization Methods Part I : Clustering Methods, Math. Prog. **39**, pp 27-56, 1987.

65. A. Rinnooy-Kan and G. Timmer, Stochastic Global Optimization Methods Part II : Multilevel Methods, Math. Prog. **39**, pp 57-78, 1987.

66. M. Bartholomew-Biggs, *Anglicised by Common Use*, Waldean Press, 1998.

67. M. Bartholomew-Biggs, *Inklings of Complicity*, Pikestaff Press, 2003.

68. J. Lucas, ed. *Take Five*, Shoestring Press, 2003.

Index